Open Wards, White Coats

Stories from a Medical Residency

Jeffrey Blake

For additional copies of this book see www.lulu.com

ISBN: 978-1-105-74468-6

Printed in the USA

Contents

Part 1: Internship

Part 2: The Residency Years

Part 3: Chief Resident

Part 1: Internship

Before the Storm

Harold Neu loved a challenge. He was the Chief of the Infectious Disease Service at Columbia Presbyterian Hospital. His face was alight and his eyes danced as I finished my case presentation to him. The patient was very sick with an infection. The cause needed to be found and the correct treatment started as soon as possible. Dr. Neu's delight was palpable as he searched for clues to the cause of an infection, a little like Sherlock on the trail of a mystery.

"Who's her doctor? We need to talk with him."

I knew who the intern was but before I could answer, the resident on the service said, "He's been paged." He was a step ahead of me, but that was okay. I was a fourth year medical student on the service as part of a clinical elective, not yet expected to be on top of everything. Back in this era at Columbia Presbyterian Hospital every patient on the medical "ward service" (meaning no private doctor) had an assigned "doctor" and this doctor was a medical intern. Although there was oversight by more senior medical residents and a "Ward Attending" most of the decisions and all of the detail work was done by a medical intern. These decisions were discussed and critiqued, but often well after the fact the next morning on "Attending Rounds." That was the Presbyterian way. That is how they made doctors out of interns and residents in their medical residency training program. The philosophy was that you learned how to make medical decisions by making them and then being held responsible.

Jimmy C had been a year ahead of me at Columbia University Medical School. He was now in the last few months of his medical internship. There was a heaviness in his movement as he approached the 9E nursing station that I did not recognize. I knew him as a happy go lucky guy from pickup basketball games in the Bard Hall gymnasium. Bard Hall was the medical student dormitory. I had made my way to the basement gym many a late afternoon or evening after classes and labs. There was nothing like a little competition and physical exertion to get the kinks out after the day's mental efforts. Jimmy was now a medical intern at Presbyterian, the main teaching hospital for the medical school. On the court he moved smoothly and had a deft touch to his jump shot. He was also a quick-witted free spirit who offered a steady stream of wisecracks during play. It was mid-April 1974, and I had not seen him on the court since he graduated over 10 months ago. Jimmy and his ward service team had requested a consultation from the hospital ID service on a 26 year old woman with systemic lupus erythematosis who had been spiking fevers. He had done multiple cultures and other tests, but so far had not found the source of the fever. Lupus, her disease, required chronic steroid treatment to keep it in remission but the steroids cause immunosuppression, making her susceptible to a number of possibly deadly infections. It was important to find the cause of the fever as soon as possible.

I had been "assigned" the case that morning by the resident on the ID service. My job was to examine the patient, review her chart, check the tests done to date, and then describe all this to Dr. Neu on attending consultation rounds that afternoon. This "presentation" had to be complete, with the focus on factors relevant to her possible infection. But it also had to be succinct. The service had many other patients to review. I was brimming with fourth year medical student importance, just two short months from my official MD diploma. I thought I knew my stuff. Little did I know how far away I was from being a real doctor. I thought my presentation was good and that I had offered a decent suggestion or two. Dr. Neu had nodded, lost in thought, but then fired off a few questions about the patient I could not answer. From experience the resident had anticipated this and paged the one guy who did know everything about this patient. That guy was the intern Jimmy C, her "Doctor".

In addition to being the Chief of Infectious Disease at Presbyterian Hospital, Dr. Harold Neu was a Professor of Microbiology and Pharmacology at the Medical school. He was a favorite among students, having taught us about the clinical side of infections as well as about the intricacies of the microbes that caused them. This fourth year elective was popular and I was fortunate to land a spot on the service in the next to last month of my medical school education. Dr. Neu was fairly young for his level of prominence in the academic world and he liked students. We liked his impish grin. His enthusiasm for his field was as contagious as many of the infections he investigated.

The national internship match had come out the previous week. I knew I would be a medical intern on this same service in six short weeks. There would be 16 of us. It was a prestigious spot. I was proud and felt fortunate to have matched there, but I was a little apprehensive about my ability to make the grade. Most of the other Columbia students who matched at Presbyterian Hospital ranked higher in the class than I did. I wasn't sure I could hold up to the rigors and tradition of the mighty Columbia Presbyterian Medical Service. Throughout the past four years there had been subtle, and at times not so subtle, reminders that medical giants walked these corridors.

Jimmy slowly lowered himself onto a stool in the nursing station as Dr. Neu talked. Our eyes met. There was fatigue all over his face – he wasn't sleepy, just weary and worn down. He answered Dr. Neu's questions matter of factly - routine stuff to him. Dr. Neu then went on to describe his concerns about a hidden abdominal abscess, the possible infectious agent, and some recommendations (mainly his, but giving me some oblique credit) on how to proceed. "What do you think, Dr. C?" Jimmy grunted. "Yeah, well, okay." He knew full well that he bore primary responsibility for deciding on these recommendations (in all likelihood he would follow them to the letter) and for carrying them out. He also knew (which I did not) that this would add time to his work day and would extend his already overflowing and never ending responsibilities in days to come.

"Okay, next case." Dr. Neu was already moving down the hall with the resident and the other medical student on the service in tow. I lingered on 9E as Jimmy pulled his patient's chart from the rack and began making notes. I was troubled by his appearance. I was hoping to reestablish some of the basketball court banter, to see a spark of humor. "So… how's it going?" I expected some crack about a sexy nurse or our past on court encounters. He just shrugged and kept writing. Without looking up he sighed and said "Jeff, I can't wait to hand this ball off to you." As I trailed the ID service down the hall toward the next patient I wondered what I had gotten myself into. If 10 months of internship at Presbyterian Hospital had knocked the humor out of this guy, what in the world would it do to me?

A First Patient

My first "rotation" or clinical assignment on the Columbia Presbyterian Medical Service in July 1974 was the Coronary Care Unit, or CCU. It was located on the center wing of the ninth floor of the hospital and often called 9 ICU. There were eight cubicles or "beds" in this unit. In contrast to the "medical ward services" where all patients were "house" patients without private doctors, the 9 ICU patients were a mixture of both private and "ward" patients. The CCU was staffed by two interns and a junior medical resident. Cardiology fellows (doctors post their residency doing specialty training) and Attending doctors supervised activities on daily "Attending Rounds" at 10 am. The two interns would be "on call" every other night, meaning they spent the night in the hospital (usually physically in the CCU itself) and were first line responders for anything and everything that happened after hours. All significant events, new admissions, transfers, patient problems were reviewed the next morning on attending rounds. For the interns this meant 28 or so straight hours of work and responsibility, then 20 or so hours off. Anything more than 2 hours of sleep during those 28 was considered a bonus.

Coronary care and intensive care units were a relatively new phenomenon in the early 1970s. Medicine had learned that patients who suffered a "heart attack" or "coronary" often died suddenly a day or two later from sudden chaotic heart arrhythmias called ventricular fibrillation. Defibrillators were devised to deliver an electric shock through the chest to disrupt this chaotic rhythm. The shock momentarily stops the heart, hopefully allowing the sinus node, our intrinsic pacemaker, to resume a normal orderly heartbeat. So "heart attack" patients were admitted to CCUs and watched with constant EKG monitoring of their heart's rhythm. The monitors were watched by nurses and were equipped with alarms so that personnel could respond quickly to ventricular fibrillation if it occurred.

So at eight am on July 1, 1974 I was one of two freshly minted "doctors" meeting with a medical resident in 9 ICU to begin learning how to be a real doctor. This medical resident himself had been a medical intern just yesterday. In academic teaching hospitals vast changes in responsibility occur every July 1. The resident described the patient in bed one, a 62 year old man admitted yesterday with an acute anterior MI. He assigned responsibility for treating him to my fellow intern. We moved to bed two. This would be my first patient. "This patient is a 76 year old man admitted two days ago with CHF. He seems stable. He's a patient of Lovejoy's." He went on to say that he needed close observation of his fluid status and adjustment of his Lasix dosage. I knew I had caught a break. At least one of my first CCU patients was "private" so I would not bear full responsibility for all the treatment. And Lovejoy was an attending I knew and liked from my medical school days. The name on the chart rack read Charles Lindbergh.

"Work rounds" were finished. We had been joined by the head nurse in the CCU for the last half of this exercise. We had a few minutes before attending rounds. I pulled Mr. Lindbergh's chart and began to read it. Could it be? Is this The Charles Lindbergh? No one had given any indication. I asked the nurse how he was doing and she said about the same as yesterday and offered nothing more. As we finished Attending Rounds Dr. Lovejoy, who as it turned out was also one of the "service attendings" in the unit for the month, said to me, "Blake, check Mr. Lindbergh tonight before you give him an extra dose of Lasix. I don't want to drive up his BUN." Treating congestive heart failure (CHF) is a delicate balance in getting excess fluid out of the lungs and not overtaxing a weakened heart and its output to the kidneys. Giving too much Lasix (generically known as furosemide) in an effort to remove lung fluid and make breathing easier can sometimes worsen kidney function. The heart in its weakened state cannot work hard enough to maintain normal kidney function if we pharmacologically deplete intravascular fluid (with Lasix) too quickly.

I went in to check Mr. Lindbergh around 9 pm. He was awake and alert but appeared tired and his breathing was somewhat labored. He had a shock of white hair and piercing blue eyes.

"Mr. Lindbergh, I'm Dr. Blake. How is your breathing? I would like to check your lungs for fluid."

"Are you a new doctor?" How did he know??

"Yes, I am a new doctor on the CCU service."

I stood transfixed.   This was The Charles Lindbergh.   I was a green intern.   What do I do now?

"What do you want me to do, Doctor?"

Mr. Charles Lindbergh had just called me "Doctor". Can I really live up to that?   I finally recovered and asked him to sit up in bed and lean forward.   With my stethoscope on his mid back I asked him to breathe in and out deeply.   There were fine crackling sounds, "rales", as he inspired.   There was significant fluid in his alveoli, the tiny air sacs of his lungs.   In normal lungs the air moves in and out with a clear blowing whoosh without extra crackles.   In CHF the weakened heart is pumping inefficiently and as a result pressure builds in the lung circulation forcing fluid to leak out into the air spaces where only CO_2 and oxygen belong.   He definitely needed a dose of Lasix.   I ordered it, but only a moderate dose.   I did not want Dr. Lovejoy to see an elevated BUN in the morning.

Mr. Lindbergh stayed in the CCU for six days. Not once did any of the staff seem to make any particular issue of his fame.   He was a patient of Presbyterian Hospital.   He was treated as all patients were to be treated – with the skill we had available and a respect for their privacy.   He was reserved and seemed to have little to say, and never demanded or expected anything special.   Dr. Lovejoy transferred him out of the CCU to a private room.   He had seemed no better or worse to me during his stay.   I did not fully understand at that time that he was dying. The treatment of CHF was fairly limited in the 1970s. I found out later that he died at his home in Hawaii in late August of that year.

In the bed next to Charles Lindbergh in that first week of my internship was an unemployed black man who was an alcoholic. He had severe high blood pressure and a recent heart attack. Both these men were treated with the same skills and compassion by the doctors and nurses. So in my first week as a "Doctor" I had made a first clinical decision on a patient who happened to be a very famous man. That has stuck with me. The unspoken lesson of Presbyterian Hospital that has also stuck with me is that whether rich and famous or unfortunate and poor, all people are due respect, care, and dignity when they are sick.

The 8 West Medical Service

The Presbyterian Hospital medical service program was organized around the architecture of the building. All medical service patients were housed on the eighth and ninth floors. Three large rectangular sections – east, center, and west extended from a central corridor. These large rectangular areas were open bedded "wards". Patient beds were separated by sliding curtains. There were no patient rooms per se. Each ward held 12 to 16 beds. There were additional 4 bedded wards off the main corridor to the east and west of the main 12 bed ward. Nurses and doctors made their "rounds" by wheeling a patient chart rack slowly around these open wards stopping at each bed to address the issues. Additional charts hung at the foot of each bed with the most recent nursing notes and "vital signs" (heart rate, blood pressure, respiratory rate, weight, urine output) on each patient. We medical interns were assigned to a different medical service each month. In my first month I was in 9 Center or 9 ICU, the coronary care unit. My second assignment was 8 West. Patients housed on the 8 West Service were all men. They were sick enough to be in the hospital but not sick enough to require intensive care or cardiac care.

There were 3 interns on each ward service and we were "on call" every third day/night. The patients on the 8 West service were divided between the 3 of us. So each patient had one primary doctor. Or more importantly to us interns, we each had 8 to 10 patients for whom we were the "doctor". Nurses were instructed to "pick up" orders written only by the intern "doctors" or their supervising residents. The "Ward Attending" doctors could not write patient orders. This was part of the way the Service enforced its philosophy of educating interns to be doctors.

New patients were admitted to the hospital each day from the emergency room and the walk in clinics on the first floor. These new admissions to the 8 West service each day and night were assigned to the medical intern on call for that day. He or she became the "doctor" for that patient for the entire time the patient remained in the hospital. Getting an "admission" meant the following. The intern received a page from the

Medical Admitting Resident in the ER. The conversation usually went as follows. "Dr. B, this is Dr. X. I got one for you. A 54 year old man with severe abdominal pain and fever. He is a drinker and his amylase is elevated. He is on his way up." Early in our internships we usually said "OK, thanks" but as the year wore on we often asked additional questions and discussed diagnostic possibilities. Some of us might even argue that the patient might not need hospitalization. That argument was self-centered because each admission meant a whole lot of work and responsibility. The argument was also futile because the decision of the MAR re hospitalization was final; and questioning the judgment of a more senior resident was not a wise move.

So our lives as new medical interns on the 8 West service quickly organized around our on call schedule. We were very busy on the on call days admitting and working up the new sick patients – usually 2 or 3 each day, and tending to the needs of our other hospitalized patients. On call days and nights were usually a blur with constant activity and decisions. The norm was 2-4 hours of sleep per on call night. There were nights we got none. The days after being on call were hazy. You stayed alert through morning attending rounds because that was when the new admissions from the day/night before were discussed in detail with the Ward Attendings. The Ward Service Attendings were seasoned doctors known for their clinical teaching ability. They all had teaching professorships at the medical school. There were two on each service for the month. As the intern on call you were required to present the case to the attending. You were to describe the history of the patient, the findings on physical examination, the laboratory findings, your differential diagnoses (what you thought the possibilities were in descending order of likelihood), the tests you ordered, and the treatment plan you had begun. It was your chance to shine, or in some cases to not look so good. Some attendings were tough and blunt. If they thought you did something wrong or stupid they would let you know. Attending rounds were over by noon. After noon most of us began a slow fade. In a daze we would catch up on what we had to do and then "sign out" or hand over responsibility for our patients to a fellow intern on call for that night. This sign out usually occurred around 5 pm. So an "on call" intern was in the hospital for 32-34 straight hours. The next day was the best of the 3 day cycle. You were fully rested. You would not get any new admissions. You could listen attentively at rounds and learn as your colleague presented new admissions and was grilled by the attending. You could attend a one hour teaching conference and not fall asleep. You could attend to the needs of your 8 or so hospitalized patients and write lucid progress notes in their charts.

The Sheriff

Some early mornings around 7 am I would see Dr. Robert Whitlock prowling around 8 West as I came up to the floor after an on call night. Even if we got some sleep we were back on the floor (hopefully after a shower) by 7 to check on patients and complete their work ups and chart notes before work rounds with the service resident at 8 am. Dr. Robert Whitlock was a professor of gastroenterology and a known expert on liver diseases. He was a University of North Carolina graduate and spoke with a southern twang. His manner was gruff and his diction was clipped. His presence oozed authority. From our medical student days we knew he was known as the Sheriff by the house staff. The moniker fit. The house staff both respected and feared him. On the mornings he appeared I wasn't quite sure why he was there. He wasn't assigned as a ward attending for the month. Why was he nosing around? If I saw him down the hall my instinct was to duck into another room. I did not want to be grilled about something I had missed or hadn't done. But he was the Sheriff. You can't run from the Law. He seemed to know about the patients admitted to 8 West the night before. "Blake, did you get a liver panel on that GI bleeder admitted last night?" Whoa, how does he know about that? Some mornings he was there, some he was not. You never knew when to expect him.

Toward the end of my first month on 8 West I had a particularly difficult night. I had 3 complicated admissions and had not gotten to bed. I had gone down to the house staff on call sleeping area for a shower around 6:30 am. I was feeling sorry for myself, but at 7 am I was back up on the floor to finish my admission note on the last patient. I was sitting in the 8 West "doctors' station", a small 8 foot by 10 foot area with phones, charting material and a small writing area. The Sheriff loomed at the door. I had a sinking feeling in my stomach – what had I missed, what had I done wrong? Is it just me, or does he check on other interns as well?

"Blake, nice work on that patient with heart failure last night. Make sure you get something to eat before rounds."

Then he was gone. Say What!? It was clipped and gruff but the Sheriff had given me a boost. Maybe I will make it as a doctor after all.

Open Wards

The wards of Presbyterian Hospital do not exist today. A large open space shared by 12 to 16 patients is not the culture of modern hospital care. But they did not seem out of place back then. In fact, the open wards had certain logic to them. The nurses from the nursing station at the corner of the open ward could observe 16 patients without having to go from room to room. If a patient had trouble he could call out or be seen by staff, not hindered by the walls of a closed room. The 8 West beds were grouped in a large open rectangular room extending to the south from the main 8th floor corridor. Eight beds were aligned along each long wall of the rectangle with a large open center area. The far or southernmost end of the rectangular room had 2 doors to an open glassed in area with comfortable chairs and small tables. It was called the Solarium and was designed for the patients who were able to get up, walk there, and sit for a spell in a nice sunny spot.

The nursing station in the corner of the large open ward had 2 glassed walls looking out to the patient beds. There was an opening to allow access, but no door. Just behind the nursing station was the small doctors' station where interns labored over their charts. It was also connected to the nursing station and the main corridor with an opening but no door. We were all in close proximity with few or no barriers. Patients, nurses and doctors were thrown together for hours on end with the patients struggling to get well and the medical staff trying to help them. In 1974 the wards of Presbyterian Hospital were not air conditioned. There were windows on the eastern side of the large rectangular open ward and these were open wide during my August rotation there. Patients and all the rest of us had to suffer with the ambient heat and humidity. It was just the way it was.

Another bit of logic to the open ward system was that the sickest patients were grouped closest to the nursing station. They could be seen and attended to more quickly. Patients would graduate to beds farther away as they improved. If a more distant patient had a relapse he would get moved back closer to the nursing station. The nurses made these decisions and they always made sense to me. In fact, early in our internships (and probably longer) the nurses knew who was the sicker patient far better than we did. On an "on call" night an intern might stand at the end of the open ward near the nursing station with the senior nurse on duty around 3 am before heading down to the second floor to catch some sleep. They both would look out at the patients and have a quick, quiet conversation about who and what to check on in the next few hours. It grew easy to get a handle on issues from that one vantage point.

I remember walking onto the ward on a morning after a night "off" and finding one of my patients moved right next to the nursing station. That was a sinking feeling. It meant that the patient had deteriorated overnight. Somehow, I had missed something. And worse, one of my "on call" colleagues had to deal with this deterioration among the new patient admissions and endless other responsibilities of a night on call. We interns, as "the doctor" for a particular patient, felt guilty when they did not do well, particularly when it happened while we were not in the hospital.

The morning routine on the wards was to get the patients freshened up and bathed between 8 and 10 am. The nurses tended and fussed over the patients. Those well enough would be gotten up into a chair next to their bed. Others not too sick or miserable would be propped up to a sitting position in their bed. At the beginning of attending rounds at 10 am many professors liked to slowly walk around the ward with interns, residents, and students in tow. They would stop to say hello to the patients and maybe briefly examine one or two to make a teaching point. The head nurse on the ward would also be along and individual nurses responsible for each patient would often stop their duties as we all stood at the foot of the bed. Nursing uniforms in those days were a crisp white. Some nurses still wore white caps. The residents and attending were in white uniforms or white lab coats. All looked neat and clean – ready to do battle against disease. There was certain pomp and ceremony to it, but it did set a tone for the students and interns. The message was that these patients were important and this was an important exercise. Looking back through the prism of nostalgia I think it conveyed a sense of dignity to the patients. They saw professors, young doctors, and nurses all expressing interest and concern in their case.

The head nurse might say "This is Mr. Jones. He had a tough night – spiked a fever to 104 but it is down to 101 now. He looks better." The intern might say "He is growing streptococcal pneumoniae in his sputum. He is on day 3 of a 10 day course of IV penicillin." The attending might have a few encouraging words for the patient. "Hang in there, Mr. Jones. The antibiotic Dr. Blake is giving you should work against your pneumonia." He then might listen to the patient's lungs and then step away saying "Mr. Jones, I want one of our medical students to hear what is going on in your lungs." He would then beckon to a student, have them listen and then question them about what they heard. "Mr. Jones, I agree with Ms. X here that there is still some consolidation in your left lower lobe. That is why you still have fever. It will get better. We will check on you tomorrow. Dr. B, how much Penicillin is he getting?" "2 million units IV, Q6H." "Good." So in this brief, 5 minute, somewhat ceremonial encounter on an open ward, several things were accomplished. A patient was buoyed, a medical student learned a small clinical fact on pneumonia, a supervising attending re-assured himself that a patient was getting appropriate treatment, and an intern got confirmation that his treatment plan still had approval.

A Patient with Pneumonia

Leroy Jones had never had much need of doctoring in his 55 years. He worked odd jobs around Washington Heights. He was not formally educated but he got by. He drank a six pack or two of beer and smoked a pack of Camels a day. In general he usually felt pretty well. In Mid-August of 1974 he began to feel tired, a little feverish, and developed a cough. The next day he coughed up some thick greenish sputum, got short of breath walking up the steps to his apartment, and felt really sick for the first time he could remember. That evening he walked the five blocks to Columbia Presbyterian Hospital outpatient medical clinic. By the time he got there he was sweating profusely but oddly he felt chilled, not hot. His breathing was labored. He knew he needed a doctor.

Presbyterian Hospital operated an open access, walk in medical clinic on its first floor. It functioned as a screening area for people with various complaints and no other access to medical care. We called it Area B. It was partly a medical clinic and partly an emergency room. For many people in the Washington Heights neighborhood, Area B was their only source of doctors. People who walked in there ranged from those with a cold or a stomach ache to those with life threatening illnesses. They would be quickly screened by a triage nurse who would get vital signs (temperature, pulse, and blood pressure) and depending on those results, the patient's complaint, and her clinical acumen, either be given a number and told to wait, or be quickly ushered into a patient cubicle where they could be examined. Area B was staffed 24 hours a day by a medical intern (PGY 1) and a first year medical resident (PGY2). A supervising attending doctor was available by phone and during very busy spells. The triage nurse took one look at Leroy Jones and knew he was sick. The temperature of 103, pulse of 115 per minute, and respirations of 24 per minute confirmed her impression. She yelled to a nurse colleague to put him in a cubicle and went to get the resident talking to a patient with stomach pain in another cubicle. "Dr. P, I have a man I want you to look at right away." Dr. P did not like being interrupted but knowing the flinty triage nurse's clinical skills he excused himself and went to examine Mr. Jones. He too, knew the patient was sick when he saw him. He noted the vital signs, asked a few quick questions, examined the patient's lungs and heart and said "Mr. Jones, I think you have pneumonia. You may need to be admitted to the hospital. I am sending you for an x-ray." He stepped out, asked the nurse to set up an IV line, and put the patient on oxygen. He then called the Medical Admitting Resident. The Medical Admitting Resident was the senior medical resident on duty in the hospital. He fielded difficult consultations on patients already in the hospital and made the final decision on any patient that got admitted to the hospital from Area B, or the medical/surgery emergency room. It was his/her decision whether a patient stayed in Area B to be quickly treated and discharged or admitted to the hospital for more extensive treatment. The MAR admitted medical service patients to the Cardiac Care Unit, the medical intensive care unit, or one of the three medical ward services depending on his assessment of the severity of the situation and the available space upstairs. He was all powerful, particularly to the new medical interns up on the ward services of the hospital. A call from the MAR at 10 pm at night meant the intern was getting a new sick patient that would involve hours of work that night, and depending on the patient, hours of work and responsibility in the days ahead.

Mr. Jones was not keen on the idea of going into the hospital but he felt awful and these folks seemed intent on helping him. He was breathing a little easier with the oxygen and had a chest x-ray taken. The two doctors who had examined him were young but seemed to know what they were doing. The last one said he was being admitted to 8 West under the care of a Dr. Blake. The next thing he knew he was on a gurney in the elevator and then in a bed in an open ward with a nurse taking his temperature and a young doctor asking him a barrage of quick questions. The next question came almost before he finished his answer to the previous one. There was an even younger "doctor" hovering around. Both these young doctors did not seem as polished as the ones he saw downstairs. Then he was being examined all over. He coughed roughly. I stopped my exam and said to the nurse "Get me a sputum culture kit". The nurse had one in her hand already and gave it to me with a look that said "Right here, Dr. B., I know the routine." I said thanks and then to the patient, "Mr. Jones. We need you to cough up some phlegm into this container. It will help us identify what type of pneumonia you have." It was not hard for Mr. Jones to comply. He hacked again and spit a large glob of thick, green material into the container. I handed it to the other young "doctor" hovering at the bedside. This other "doctor" was actually a third year medical student doing his first clinical rotation and had been assigned to the 8 West service. His assignment this night was to shadow me and talk to the same patients I did. He would be responsible for "writing up" a patient who was admitted and "presenting" this summary to his preceptor in the morning. I said "we are going to do a gram stain on that sputum before sending it to the lab for culture. Take it down to the lab and get the gram stain material set up. Make sure to keep the specimen sterile."

The eighth floor of the medical service had a small laboratory with microscopes, glass slides, and chemicals for fixing and coloring materials. We, the medical house staff, used this lab whenever we wanted to do something with medical specimens. There was equipment there but no technical staff. Part of the Presbyterian philosophy was that doctors needed to check things for themselves before making clinical decisions. We examined blood and urine specimens as well as sputum in that lab.

My heart rate picked up a notch as I focused the microscope. Damn, those are gram positive diplococci. There were pairs of fat, red colored bacteria inside a white blood cell which had been fixated on the glass slide of Mr. Jones' sputum. I had carefully put a sterile swab into the sputum culture container, extracted a small amount, smeared it on a glass slide, and then applied the staining chemicals in sequence to the slide. When we have infections our bodies send special soldier - like white blood cells to the site to fight the infection. These white blood cells (phagocytes) eat and enclose invading bacteria if they find them. Seeing such bacteria inside a white blood cell on a freshly stained specimen usually gave a good clue regarding the cause of the infection. A patient with gram positive diplococci in their sputum almost certainly had pneumococcal pneumonia. The pneumococcus is a bacillus of the streptococcal family and exists in pairs locked together. It stains red (positive) when fixed by the gram staining chemicals. In the 1970s it was the most common bacterial cause of pneumonia. It could be deadly, particularly if it involved a whole lobe of a lung. Our microbiology teachers and later Dr. Harold Neu had drilled all of this into us.

"He's got gram positive diplococci in pairs!" I said to the medical student and to the room in general.

I was excited because I was 6 weeks into my internship and I had made a specific diagnosis. At least I thought I had. I told the medical student to have a look and went to find the house resident on call for the night. I wanted confirmation of what I saw. It was my first patient with pneumonia.

The resident confirmed my analysis. "Wow, this specimen is loaded with pneumococci! Have you started treating him yet?" "No, I wanted to get all the data and present it to you, first." That was the usual protocol. The last thing I wanted to do early in my internship was to start a treatment that was incorrect without checking with a more senior resident. "Yeah, I know, but from what you say about the patient and looking at this sputum the sooner we treat the better. What do you want to start him on?" This I had remembered from Dr Neu's teaching so I said, "Pen G, a million units q6h." He told me to get on it, complete my workup and then page him to present the whole case to him. I hurried back to 8 West, wrote the order in the nursing order book, and then spoke to the nurse about the Penicillin. She said fine and also informed me that she had gotten a urine specimen in two tubes, one to send to the lab for culture and one for me to look at myself in our lab. She asked if the oxygen at 2 liters per minute was ok and if I still wanted the IV to run at 150 cc per hour. She also had the EKG machine at the bedside. She knew that we would need and were expected to have an EKG on a patient admitted to the hospital. Thank god for nurses. The good ones made our lives as interns possible. They thought of things before we did and often made it seem like we had thought of them ourselves. So in working up Mr. Jones as a new admission to the hospital I had taken a complete medical history and done a thorough physical examination. I had checked the stool from the rectal exam for occult blood. I had looked at his chest x-ray in the radiology department on the fourth floor and seen the white cloudiness in the left lower lung field. I had looked at his urine under a microscope and at a smear of a drop of his blood, in addition to looking at the gram positive diplococci in his sputum. I had written his admission orders in the nursing order book and discussed them with the nurse. I had shared some thoughts and reviewed the specimens with the medical student on call with me. Mr. Jones had arrived on 8 West at 10 pm. He had gotten his first dose of IV penicillin before 11 pm. It was approaching midnight by the time I had paged the house resident to present all the data to him. Usually the medical student would join us for this discussion. It was informal and succinct but a good resident would make an attempt to do some teaching to the student as well as the intern. That was part of the Presbyterian philosophy as well. We learned from those more senior to us. They were expected to teach us as well as make sure we did the correct thing. It was a preview to the more formal and detailed attending rounds to take place the next morning. Both the intern and the resident wanted to get it right and not miss something. If there was criticism by the attending it would be directed at the intern, the "doctor" on the case, but the resident knew he was responsible for the decisions as well. After 15 minutes of review and consideration of alternatives and other issues on Mr. Jones (his over use of alcohol, his enlarged liver) both the resident and I were satisfied with our plan. He moved on to go see another case admitted to another intern on a different service and I went to write up my admission note in Mr. Jones' chart. If I did not get another admission from the MAR I might get to bed by 2 am.

By 2 am Mr. Jones finally fell into a fitful sleep. He had been questioned, prodded, examined, moved around, and stuck with needles for over 3 hours. After I finished examining him, the nurses had to finish their work and give the IV medication. Then the medical student went back to ask him more questions and examine him again. Mr. Jones was tired, sick and just wanted to sleep. But medical students were expected to do their own history and physical on any patient they presented to their preceptors so Mr. Jones had to go through this exercise as a patient on the ward service. I remembered feeling badly for patients when I did this exact same thing when I was a medical student. But that is how med students and doctors learn. Sometimes patients do go through a lot.

When I checked him he was pretty much the same. His temp was still 103, his left lung still sounded lousy. He was still breathing faster than normal. I told him we had started treatment that should help his pneumonia and that we would let him rest now. As I went downstairs to try to catch a little sleep I was thinking "Wow, I have just diagnosed and begun treatment on a patient with a severe pneumococcal pneumonia. I wonder if he will be better in the morning."

Leroy Jones was a little better the next morning. His temperature was down to 101, he had slept a few hours and he felt somewhat better. But he was still coughing and his breathing wasn't right. He reckoned if he had the pneumonia the hospital was a good place to be.

Attending Rounds

"This is the first CPMC (Columbia Presbyterian Medical Center) admission for this 58 year old repairman admitted with the Chief Complaint of fever and cough." We were in the 8 West Solarium and I was presenting the case to the Ward Service Attending. My two fellow 8 West interns, the 8 West resident (PGY2), 3 medical students, and the 8 West head nurse were grouped in a circle of chairs. I went through the progression of his symptoms, the fact that no one else in his family was sick, his past medical history (there wasn't much), and his life style habits. I was about to begin my description of the physical exam when the attending stopped me with a question. He directed it to one of the medical students, the one who was on call with me last night. "Is there anything else you might want to ask in the history?" The student floundered a bit but recovered and said "Has he traveled outside the country recently?" thinking of possible exposure to exotic infections. "Well, good thought, but not what I was after. It doesn't sound like Mr. Jones has the means to do much traveling." Looking at me he went on "Dr. Blake, has he been exposed to anyone with TB, - family, coworkers?" Luckily, I had asked that question last night. "The patient denied known exposure to anyone with recent TB or a known respiratory illness." I silently thanked the teaching of Dr. Neu. There was a fair amount of tuberculosis in the Washington Heights area. I knew that this patient's presentation and his chest x-ray were not suggestive of TB but the attending and the others did not know that just yet. The attending was making the point to all of us to always consider tuberculosis (TB) in a patient with a fever and a cough. To me he was making the point of including that thought (and possible exposure) when I presented the history.

I described my physical examination starting with the vital signs, emphasizing the positive lung findings, noting the enlarged liver, and mentioning the pertinent negatives. These are statements confirming the lack of abnormality on a part of the exam that could be present with other causes of a febrile illness. Pertinent negatives let the listener know you thought of a possible diagnosis associated with certain findings on exam but that they were not present. The attending again stopped me as I was about to review the laboratory findings. He addressed another medical student. "What are you thinking so far? A patient with fever, productive cough, and signs of consolidation on exam probably has pneumonia. What do you want to do now?" A decent student would reply that he would get a chest x-ray and do a gram stain on his sputum. The attending nodded for me to continue. "His CBC showed an elevated WBC count with a shift to the left. Sed rate was 42, electrolytes were normal, urinalysis and EKG were normal. We have his chest x-ray here. Bony structures are normal, heart size is normal. There is a large infiltrate in the left lower lobe. No air fluid levels are noted. The sputum gram stain showed many WBCs and gram positive intracellular diplococci in pairs." Nods of recognition spread around the room.

The attending then asked a fellow intern how he would treat this patient. After the reply he said to me "Dr. Blake is that the dose of Penicillin you used?" It was and he agreed that it was correct. He asked me what else I thought about the patient and other possible evaluation that I thought might be needed. I described my concern about his overuse of alcohol and his enlarged liver. I reported that liver function studies were pending and that we would consider scanning his liver. Finally, he asked the resident on the service to list the differential diagnoses for community acquired pneumonia to make sure we all at least thought of the other possible causes of pneumonia. Not all are as straightforward as this patient with pneumococcal infection. As we finished I realized the attending had gotten us all involved in the discussion and gone up the ladder of experience by asking the easier questions of the med students and the more difficult of the service resident. If we all paid attention we all learned something. That was the function of attending rounds.

Group Clinic

Mr. Jones' hospital course was a little rough the first three days. He continued to cough and spike fevers to 103 but after his fourth day of IV penicillin all parameters began to improve. His temperature hovered around 99-100 and his cough gradually diminished. He began to breath comfortably even without oxygen. In the current medical era he would have been sent home to complete his 10 day course of IV medication with the aid of home nursing visits. But in 1974 we kept him in the hospital for the full 10 days of IV medication and watched him an additional 24 hours off medication before discharging him. We wanted to be sure he did not relapse off the penicillin. Home health visits by nurses were not available back then. In addition, in the hospital we could keep him away from smoking cigarettes and drinking alcohol, two issues that probably contributed to his susceptibility to pneumonia. His liver chemistries were elevated and his liver scan consistent with early cirrhosis from the alcohol. By discharge the liver abnormalities had improved somewhat. We did no more about the liver issue other than lecture him about the negative effects of alcohol.

On discharge Mr. Jones was given an appointment for follow up in the Medical Service Outpatient Clinic. We called it Group Clinic. I am not sure why but I think it was because the patients we saw became part of our "group" practice. All patients who did not have regular doctors (and all the ward service patients did not) were followed in this clinic after hospital discharge. The clinic was staffed by medical interns and residents. Each house officer had a clinic session one afternoon a week. Mr. Jones had not been a Presbyterian patient before, so on discharge he was sent to my group clinic. I was his "doctor" in the hospital and would be his "doctor" as long as he remained a clinic patient. The idea was that patients were better cared for by doctors who knew the most about them and that interns and residents ("doctors") learned best about disease process and patients by following them throughout the course of their illnesses both in and out of the hospital. It was known as continuity of care. The concept was good but the problem was that we had very little training about outpatient care in medical school. We learned on the fly.

Throughout my internship year and the additional 3 years I was part of the Columbia Presbyterian Medical house staff I saw outpatients on Tuesday afternoons. We rarely had to see more than 3 or 4 patients but it was hard to go down to an afternoon clinic after a night on call and morning attending rounds.

All we wanted to do was finish our work upstairs and get out of the hospital and home to bed. The good thing about Group Clinic is that we got to know the patients. I remember bonding with some and was saddened to leave them when I moved on. The theory of group clinic was good. Looking back I realize its execution was not. We had not been taught much of anything about outpatient care in medical school. We were well supervised by senior residents and attending physicians in the hospital setting but not in clinic. I remember feeling my way by relying on nurses and other interns or residents who happened to be in the clinic when I had questions or concerns. There was a medical attending assigned to the clinic to supervise but I do not remember them being that attentive. Current medical school curricula and residency programs put a lot more emphasis on outpatient care. In my Presbyterian era the excitement was in-patient medical care. "Real doctors" showed their stuff on sick in-patients with life threatening diseases.

I did see Mr. Leroy Jones 2 weeks after his hospital discharge. He had not had any fever or cough and his lungs sounded fine. I again cautioned him about alcohol consumption but not knowing what else to do I simply scheduled him to come back for a check-up in 6 months. I had no specific plan as to what should be done. Nowadays various preventive measures like serial blood pressure checks, lipid analyses, immunizations, nutrition counseling, and cancer screening would be emphasized. Mr. Jones never did

come back and to my knowledge was never readmitted
to the hospital during my tenure. One hopes he did well.

Eight ICU

Other than emergency rooms the real hot spots in
hospitals are intensive care units. The sickest patients
end up there and dramatic "saves" do occur. When I
compare the technology of what occurs now with what
we did back in 1974 I am amazed that we saved anyone.
But we did, and one of those saved was a man named
William W. Mr. W had never been in a hospital, and
like many patients we saw back then, never had much in
the way of outpatient care before he landed in a hospital
emergency room. He had smoked two packs of
cigarettes per day for 40 years, starting as a teenager.
He did see a neighborhood doctor a year before his
hospitalization who told him he probably had
emphysema and to stop smoking. Like the great
majority told to stop he did not and the damage to his
lungs was already far along. Mr. W looked sick when
he arrived in the Presbyterian Area B emergency walk in
area. He was so severely short of breath that the triage
nurse bypassed all paper work, got him on a stretcher
sitting bolt upright, and slapped an oxygen mask on him.
The medical admitting resident took one look, quickly
listened to the lungs, ordered an arterial blood gas
measurement, a heart monitor and dialed the number for
8 ICU. "58 year old man with COPD in severe

respiratory distress. Probably needs to be intubated. He is on his way up."

8 ICU was located off the same central eighth floor hallway as were the 8 West and 8 East wards. Like the 8 West ward service, 8 ICU had no rooms or cubicles, just 8 beds arranged with four along each wall in the large open rectangular space. There were screening curtains available to give some privacy but these were rarely used. Most patients were unconscious or too sick to care about privacy. It was not a peaceful, quiet hospital setting. The place crackled with intensity and activity. We, the medical house staff, had the sense that this was our battleground, a place where we were truly tested. This was where we made our stand against the ever present enemy of doctors everywhere – sickness and death. The sound of ventilators whooshing and monitors beeping was a constant background. The air seemed to carry a vague odor of sickness and decay. There was often a flurry of activity around one or more of the beds with nurses and doctors tending to the latest and most acute crisis. For an intern in 1974 a first rotation in 8 ICU was an exciting challenge but also very intimidating. These people were really sick. Every action and decision was crucial. Like in the cardiac intensive care unit, we interns worked shifts of 24 hours on and 24 off for 3 - 4 straight weeks. The 24 off was really only 20 because you did not leave the unit until after attending rounds each morning at 10 am. The unit was run by the second year medical resident who supervised the 2 interns. He and the unit director, Dr. Glenda Garvey, really made the majority of the

management decisions. We interns mainly implemented
them. We also did all the heavy lifting, the "scut" work;
drawing blood, taking fluid cultures, adjusting ventilator
settings, doing gram stains and microscopic analysis of
specimens, running to radiology to get x-rays, calling for
lab reports, and writing daily progress notes and orders
in patient charts. We intern/doctors were made to feel
that everything about our patients was our responsibility.
I recall being absolutely fearful on attending rounds
during my first ICU rotation that I would appear stupid
or not know something about the intricacies of the
patient's dire state. There was so much data on these
very sick people and that data was constantly changing.
We were expected to know it all.

In the present era there are "intensivists", doctors
who train for two years specifically for work in ICUs
after completing a medical residency. These intensivists
man ICU units 24 hours a day. Today's medical
residents do work in these units, but under the
supervision of these trained ICU specialists. In the
1970s we medical residents ran 8 ICU pretty much on
our own. There were daily attending rounds to review
the cases each morning but these dealt with after the fact
decisions. The critical minute to minute, life and death
decisions were made by doctors one and two years out
of medical school. It was both exhilarating and
terrifying. There is no better way to learn how to make
critical decisions than to have the responsibility for
making them. Just as a surgeon learns a procedure by
making the incision and doing it on live patients, we
medical residents learned critical care by doing it on live

patients. Were mistakes made? – Undoubtedly some were. Were some patients harmed? – Probably a few. But most were helped and some even "saved". And we medical house officers at Presbyterian Hospital learned critical decision making that stood us in good stead throughout our medical careers.

The gurney with Mr. W burst through the swinging doors of 8 ICU with an ER nurse on one side and the medical admitting resident on the other. I and two ICU nurses stood waiting. The group of us manually lifted Mr. W onto an ICU bed. There was an art to this. Two of us stationed at the head of the gurney, two at the foot. We all grabbed a corner of sheet under the patient and "one, two, three!" lifted him, sheets and all, from the gurney to the bed. The head of the bed was cranked up so he could breathe easier, monitor leads were attached, IV tubing was switched from gurney to hospital bed, and a blood pressure cuff was wrapped around an arm. All of this took no more than 30 seconds. The movements were quick, intense, orderly and efficient. The nurses introduced themselves and calmly told Mr. W that they would take good care of him. I too, introduced myself, and as the "doctor" had first crack at examining him. I had seen sick patients in medical school but never this acutely ill and as yet untreated; and never with the awesome responsibility to do something about it. Mr. W was gasping for air despite the oxygen mask on his face. His eyes were wide with terror. His heart rate was 140 (normal less than 100), his lips and fingernails cyanotic (blue), and his chest full of wheezes. The second year medical resident supervising the unit was hovering over my shoulder. I said "I think he needs to be intubated." We both knew the results of the arterial blood gases drawn in the ER. "Page anesthesia STAT" he said to one of the nurses. He had been waiting for me to make the decision on intubation and probably would have given me another 30 seconds before the stat call to anesthesia. Mr. W's ABG (arterial blood gas) showed very low oxygen and very high carbon dioxide levels in his arterial blood. His lungs were so damaged by the emphysema that he did not have the necessary cell structure to get O_2 into his blood and CO_2 out. He would need the help of a machine, a respirator, to assist his breathing. Without more O_2 and with the poisoning effects of too much CO_2 in his blood his cells would begin to shut down and he would die.

Intubation means putting an endotracheal tube (curved plastic tubing about 8 inches long and with a hollow center channel to carry air) down through the mouth, between the vocal cords and into the trachea. The trachea is the anatomical tube connecting both lungs to our throats and the outside air. The trick is to get the endotracheal tube into the trachea and not into the esophagus which lies just adjacent. If you pump air into the esophagus instead of the trachea the stomach blows up and the patient gets less, not more oxygen. A patient has to be intubated with an endotracheal tube in order to use a mechanical ventilator. You cannot pump oxygen under pressure through a mask in the concentration and frequency needed for patients whose lung function is failing. They need more O_2 forced into their lungs and a larger volume of air with each breath. The larger volume of air also allows them to exchange the large amount of poisonous CO2 that has built up.

It is very hard to see where to put the tube, particularly in a patient who is terrified and gasping for breath. A laryngoscope is used to open the mouth, extend the patients neck and shine a light down the throat so that the vocal cords can be visualized. If you get the tube through the cords you are into the trachea, not the esophagus. Anesthesiologists intubate patients every day in the operating room before surgical procedures requiring general anesthesia. They are the ones in a hospital most skilled at this maneuver and are always paged to do this in a critical setting. We medical residents and interns had practiced this procedure on cadavers but had done few, if any, on live patients. In the OR patients are sedated and partially asleep with anesthesia before a tube is inserted. It is much more difficult in an emergency setting. You cannot sedate or anesthetize the patient because that may worsen lung function. Worse yet, the last thing an awake, air starved patient wants is a large plastic tube shoved down their throat.

Mr. W was not going to last much longer. His heart rate stayed high and his O2 sats were dropping, even breathing the high concentration of oxygen through the mask. The anesthesia resident had arrived and was getting ready. The intubation instrument tray was open next to the bed. The nurses had pulled the emergency crash cart with all the emergency meds and equipment next to the bed, just in case. I was at the patient's side trying to explain to him how bad his lung function was and what we were planning to do. Who knows how much he understood. A nurse was on the other side of Mr. W's bed opposite of me, the anesthesiologist at the head of the bed, and the ICU resident standing just over my shoulder observing the heart rate and O2 sat monitors. I looked at the anesthesiologist. "Ready?" He nodded. Did I detect a slight bit of unease in his eyes? How many of these emergency intubations could he have done? Mr. W. was sitting upright in the bed. When you can't breathe and your lungs are shot the natural urge is to sit up, not lie down. We cranked the bed down flat and the nurse and I held down the patient's shoulders and chest. To get an endotracheal tube in, the patient has to be flat so that you can see the vocal cords and trachea through the laryngoscope. The anesthesiologist extended the patient's neck by tilting his head back, threw off the O2 mask, pried open the mouth and inserted his light. Another nurse held the patient's head in place. Mr. W. thrashed a little but seemed to understand he had to hold still for this procedure to work. I think it was the soothing words being murmured in his ear by the nurse at his side. We all knew we had one chance at this. If that tube was not in the trachea pretty soon Mr. W. would stop breathing all together or his heart would fibrillate. I looked at the anesthesiology resident inches from my face as I held Mr. W.'s right shoulder down. His concentration was fierce but sweat beaded all over his forehead and upper lip. The seconds ticked by and then "I'm in!" One

nurse grabbed an ambu bag, attached it to the
endotracheal tube and hand pumped air down the tube.
The ICU resident put his stethoscope over the left lung,
and then the right as the nurse rhythmically pumped the
ambu bag. "Yep, good air flow". If the tube had been
in the esophagus he would not have heard the whoosh of
air. The nurses secured the endotracheal tube with
adhesive tape so it would not move. Mr. W's O2 sats
began to climb slowly and his heart rate slowed a bit.
The nurse and I relaxed our grip on Mr. W. The
anesthesia resident stood back visibly relieved. We
cranked Mr. W's bed up slightly. Our own heart rates
also began to slow a little. Disaster had been averted, at
least for now. There was an air of satisfaction around
that bed. Five young medical personnel knew they had
done a pretty good job on a critically ill patient.

Complications Occur

But that was just the beginning. Mr. W and I
had a few more adventures together in the next 3 weeks.
That night we sedated him with a dose of intravenous
medication. Now that we had control of his breathing
with the ventilator we could put him to sleep and let the
machine breath for him. We got a portable chest x-ray
to check the position of the endotracheal tube. The
distal tip cannot be down too far or one of the main stem
bronchi to the right or left lung would be partially
blocked. The x-ray confirmed good tube position but
also showed a hazy infiltrate on his right lower lung

field. I obtained an arterial blood gas after he had been on the ventilator for one hour to see if his arterial blood showed higher O2 and lower CO2 concentrations. It did. We were going in the right direction. We needed to know arterial, not venous, gas concentrations because arterial blood more accurately reflects how the lungs are performing. The arteries are deeper under the skin than veins and you cannot visualize them as you do veins. They have to be palpated and the needle inserted straight down into them with the angle of insertion at 90 degrees to the skin. They are drawn frequently on patients on mechanical ventilators to assess pulmonary status. We interns did a lot of ABGs in 8 ICU and soon became quite proficient.

The nurses were able to insert suction tubes down the endotracheal tube and suck out accumulated secretions. They got some brownish green, thick sputum from Mr. W's lungs. We sent it for culture. I looked at it under the microscope and saw many white blood cells and a hodgepodge of blue or gram negative stained bacteria. He did not have the classic gram stain findings of pneumococcal pneumonia as Mr. Jones from 8 West had. But we figured he had some type of pneumonia because of the thick sputum and the x-ray findings. Patients with emphysema who decompensate this severely usually have an underlying infection as the precipitating cause. We started him on an antibiotic that would cover several possible types of bacteria. We could change to a more specific antibiotic when the exact culture results of the sputum came back in 24-48 hours.

Mr. W spent 10 days on the mechanical ventilator. We gradually reduced how much work the machine did for his lungs and gradually decreased his sedation so that his own respiratory drive would kick in. He was more alert and began to fight the tube taped into his mouth and throat. He constantly had to be reminded not to attempt to pull at it. He had completed 10 days of an IV antibiotic and his chest x-ray showed improvement. He began to tolerate a few hours of breathing through the tube on his own with the machine supplying just oxygen, not pumping air into him. Once a patient has been assisted by a ventilator for over a week the body forgets how to breathe on its own to a certain degree. The lungs and central respiratory drive center have to relearn their roles. We had reached a critical decision point. Was it time to pull the endotracheal tube? Was his lung function improved enough to fly on its own? Mr. W made very clear signals that he wanted the tube out. It was hard to blame him. Besides the discomfort he could not talk or eat. But the last thing we wanted was to pull the tube too soon and have to put it back in a few hours.

The first 24 hours went fine. He was hoarse from the effects of the tube on his vocal cords but Mr. W loved to talk. I found him quite the character. We began to think about transferring him out of the ICU to be watched on 8 West for a few days before a possible hospital discharge. But around 10 pm the second day post extubation his heart rate started to climb a little and his O2 sats fell just slightly. The nurses called me over. He had lost a little of the light in his eyes and there was no humor in answer to my questions. I ordered some aerosolized bronchodilating medication and went to check another patient. He got the medication but 30 minutes later was clearly laboring and wheezing. His sats were getting dangerously low. I knew he had to be re intubated. So the process of 12 days before was repeated. Fortunately, it again went rather smoothly. But we were all disappointed. Mr. W had become a favorite of all of us. We knew that the longer a patient spends on a mechanical ventilator the more difficult and less likely it is to get them off.

As often happens in the ICU setting, a patient you think may have turned the corner to become a "true save" has a sudden complication and backslides back into a dire situation. It is the nature of the beast. But some patients seem to have a little more spunk or maybe a little better physiology. They continue to make comebacks from the edge. Mr. W was one of those. He was back on the respirator and that was not good, but he again made rapid improvement. This last decompensation/complication was an episode of bronchospasm, not further deterioration of his basic lung function. Bronchospasm (constriction of the airways like in asthma) can be reversed with medication. In two days he had improved enough to be extubated. So again, it looked like Mr. W. was going to be a save but again the ICU medicine gods said, not so fast. It was near midnight and fairly quiet in the unit when Mr. W went into a rapid, potentially fatal heart arrhythmia, ventricular tachycardia. Monitor alarms went off, nurses yelled, I jumped. I was able to "shock" him out of the V

Tach with the external defibrillator. It was the first time I had done that to a live patient. It worked. One more complication dealt with. That son of a gun had 9 lives. He improved enough to be sent out of the ICU in 3 days and discharged from the hospital in another week. He spent most of the next year out of the hospital. But his bad lungs and smoking finally caused his death on another ICU hospitalization late in my junior residency year. I knew I had helped him and been a part of his survival during his initial hospitalization. He really had been a save. But as often the case in medicine, he, the patient, had also helped me. I learned a lot about ICU medicine and the care of respiratory failure because of him. I also learned that some people have a lot of courage and dignity in very dire medical situations.

The Drinker

The two iron lungs sat along the hallway wall between 8 ICU and 8 West. The cylindrical metal tanks looked cold and intimidating to me. I had been walking past them during the first months of my internship and was not sure why they were there. I knew they had been used in the 1950s for patients with polio but that is all I knew. We had not been taught their utility in medical school. They had been replaced by modern, positive pressure mechanical ventilators. Maybe these two iron lungs were historical relics waiting to be sent to a medical museum.

My beeper sounded with the page to call the Medical Admitting Resident in the ER. This would be an admission – a new patient to work up and treat. My heart rate picked up a little and I got a sinking feeling in my stomach – a mixture of excitement over the challenge and anxiety as to whether I could handle it. The MAR gave a clipped bulletin on the patient.

"Got Mr. C down here. Chronic COPD. He's laboring a bit. Needs a night or two in the Drinker."

"The What??"

"The Drinker… the iron lung. He comes in for a tune up every two months or so."

It was impolitic for an intern to ask too many questions of the MAR – he was two years my senior and busy with many other triage decisions. But I had no idea how to get a patient into an iron lung. My anxiety level increased – a common phenomenon during internship. I was a doctor but there was all this stuff I did not know. How would I ever learn it all? Dreading the feeling of stupidity, I paged the House Resident, one year my senior and my direct and only supervisor for the activities of this night on call.

"Yeah, he comes in all the time. Hardly ever needs to be intubated. An easy admission really. Just get an ABG, stick him in the Drinker, and check an ABG later tonight."

I had several questions, one being why in the world is it called the Drinker? But all I mustered was to ask how to set it up. The response was not to worry. There are no settings. Just get the nurses to get him in it and turn it on.

Mr. C was in his sixties and had the barrel chest and raspy voice of a long time smoker. He was sitting upright on a stretcher next to one of the iron lungs. He was laboring with his breathing but in no real distress, had stable vital signs, and no fever. He said his breathing had worsened the last few days and he thought he needed a rest – meaning a night in the drinker. He had done this numerous times. I briefly examined him and then got the arterial blood gas specimen from his right radial artery. The blood in the tube was a dark purple. Well oxygenated arterial blood is bright red. It was clear his lungs were not functioning well. Would this iron lung fix him? He seemed to think so. He seemed almost jaunty as he scooted over from the stretcher into the opened cylinder – like he was meeting an old friend. The night nurse, Diane, had opened and prepared it for Mr. C's night. The darn thing opens longitudinally along its midline – sort of like opening a coffin. When Diane closed the top he was entombed, except for his head. There is a soft, pliable plastic material that forms a collar around the neck and maintains a seal so that the pressure changes generated on the inside of the iron lung are maintained. Diane asked if he was all right and ready to go. "Fire it up, Diane" was the response. The guy was on a first name basis with the nurses. I felt a bit like a fifth wheel. Diane flipped the switch and the iron lung started whooshing and clanking – pulling and pushing different pressure levels inside that would move Mr. C's chest cavity in and out and make his effort of breathing easier. I was fascinated by the process and a bit dumbfounded.

"Does he stay in there all night?"

"He usually falls asleep. We'll get him out for breakfast. So, Dr. B, when do you want that follow up ABG? Two am?"

It was 11 pm and 3 hours seemed a good period after which to assess his status. I called my House Resident and told him we had Mr. C in the Drinker and that his baseline ABG looked pretty bad with a low O2 and a high CO2.

"Do I take him out of there to do the follow up ABG?"

"No, just use the arm holes."

There were two circular holes about four inches in diameter along the side of the machine. These holes were sealed with the same collar- like pliable plastic seal that surrounded the neck. You could stick your hands through. The collar sealed around your arms so that the lower pressure inside the cylinder was not changed. So at 2 am I came back to the Drinker to recheck the ABG. I was pretty good at the technique but had never done one inside a moving cylinder. I put my hands through the seals with the alcohol swab and the syringe loaded with a #25 gauge needle. It was a strange sensation. My hands moved a few millimeters with each whoosh, clank, and pull of the machine. This was going to take some fine timing. Mr. C sensed my hesitation.

"Just stick me between breaths, Doc. Try not to miss."

I understood that. Radial artery sticks hurt. The needle often hits the surface of the radius bone just beneath the artery. I wiped the skin with the alcohol swab, moved his arm a little closer to me so that I had a more comfortable angle to push in the needle. I had my chest against the outside of the cylinder trying to steady myself. Diane was standing nearby with the lab slip and a cup of ice in which to transport the syringe to the lab. I had about ten seconds of stillness between negative pressure movements of the iron lung and my fingers. I pushed the needle in, held the syringe in place with my left hand and drew back with my right. Bingo! Blood surged into the syringe. It wasn't bright red but was redder than the dark purple color of his first ABG. He was improving. I pulled the needle out and applied pressure to the artery with my left hand as I held the syringe in my right. After the next whoosh I pulled the syringe out of the cylinder arm hole and handed it to Diane. I kept my left hand on his artery. You had to apply pressure for five minutes or so after an arterial stick to prevent bleeding. Diane looked at the syringe with the reddish blood, gave me a quick glance and said

"Nice."

I wasn't sure that she meant nice work on my part or nice that Mr. C's oxygenation was improved. It was probably the latter, but I took it as both. I told Mr. C that his O2 sat looked better. As I held pressure on his artery he seemed to drift off toward sleep. When I pulled my hands out of the drinker a few minutes later he stirred and opened his eyes.

"Night Doc." And then he winked at me and said …"Nice."

During my internship and residency years at CPMC I saw and treated a few more patients in these iron lungs. They were quite effective in helping patients with COPD and emphysema if their condition was not too badly decompensated. It wasn't until several years later that I learned why these old relics were called Drinkers. In 1927 the iron lung was invented by two researchers at Harvard. The prototype consisted of an iron box and two vacuum cleaners. One of the two inventers was named Phillip Drinker.

Things Get a Little Strange

It was one of the better rotations during my internship year. We called it the Area B Graveyard Shift. It was a 3 week stint in the combined ER/Walk - In Clinic and the hours were 10 pm until 8 am. The Columbia Presbyterian Hospital Medical Center complex covered a large 4 square block area in the upper Manhattan neighborhood of Washington Heights. Area residents were generally poor and a mix of Hispanic, Black, and White. The poverty gave the neighborhood a grimy, rough edge. We interns and residents were the docs for these people. Their portal for entry to medical care was to walk into Area B any time day or night.

I liked the rotation because in contrast to most internship work it had a finite time responsibility. At 8 am you handed off any remaining issues to the intern and resident coming on for the day shift. You went home to sleep. I also liked the way we learned to make triage decisions. Obvious trauma and people in medical distress were sent to surgical and medical residents by savvy triage nurses for quick treatment and possible hospital admission. Patients judged less sick but in need of seeing a doctor were shunted to us medical interns. Our "office" was a cubicle about 4 feet by 4 feet which opened on 2 sides. Side one was open to the patient waiting area with rows of hard plastic chairs. Those chairs were usually full of waiting patients. The other side was open to a hallway and the staff work area. The cubicles could be closed for "privacy" with thin pull - curtains. Our job was to decide what was wrong with the patient if we could, but more importantly to decide if they were sick enough to merit evaluation by the Medical Admitting Resident for possible admission to the hospital. That was a fine line to walk. You did not want to dismiss patients and send them home after a brief interview and quick exam if they really had a serious issue. But you also did not want to call the admitting resident for every patient who walked in with a headache, sore throat, or stomach ache. We learned to use a brief history, vital signs, a targeted exam, and stat lab tests to figure out if a patient was really sick, ill with a nonthreatening short term issue, or just part of the worried well. In Area B the patients we saw ran this whole spectrum.

As best I can tell the graveyard shift is a general term for late night shift work or sentry/guard duty. The implication is that things are generally quiet but one has to stay alert for possible strange happenings. Not so in Area B - at least the quiet part. At 10 pm and later it was usually full of patients of all ages with varying levels of distress. I was often perplexed, amused and frustrated that people with the "dizzy weaks" or a stomach ache for 2 days would wait until 2 am at night to come in to be evaluated. The influx of new patients usually slowed to a trickle by 4 am. And by that time the sicker people had been admitted and the less sick treated, reassured and sent home. The pace often picked up again around 6 am with early risers. That 4 am lull was looked forward to by the whole staff, not just us docs. In that lull we would often sit down in a small, dingy lounge for some snacks and shared stories. Occasionally a nurse would bring in some cookies or a cake - a definite upgrade from the usual hospital fare of chips, pretzels, and vending machine ring dings.

Toward the end of my 3 week graveyard shift rotation I drifted into the lounge around 4 am. I was weary and looking forward to sitting down for a few minutes. It had been the usual hectic night. We still had an asthmatic young woman in one of the back rooms whose wheezing had been pretty severe when she came in at 2 am. She had cleared somewhat with bronchodilator treatment but she needed reevaluation. And I needed to decide whether or not to have the admitting resident see her. It was otherwise very quiet. The waiting area was empty. Ted was in the lounge with 2 of the nurses. He was a nurses' aide and helped out with odds and ends. He was on the spacey side and his persona was similar to some of the late night characters that visited Area B. But his calm demeanor and helpful attitude were appreciated by all. That night he had brought in a tin of brownies, taking his turn to supply our late night snack. We all sampled one or two. I was the only doc still in the area. The surgical residents and my senior medical resident had gone up to the house staff quarters to catch a little sleep. We had a few laughs and then Martha, the Area B head nurse, reminded me that I needed to check out the young asthmatic patient before we got busy again. I felt a little strange as I walked in to examine her. When you listen with a stethoscope to a patient whose airways are constricted the air movement makes high pitched musical sounds – wheezes. As I listened to this patient I found myself kind of grooving to the musical sound of her wheezing. Whoa!! What is this!? What am I doing? I was not thinking about her condition or what more needed to be done. Something was wrong and I knew what it was. Being a product of the late 1960s I had a degree of familiarity with the sensation of marijuana high. This seemed awfully like that. It had to have been the brownies. To some degree I was amused, but this was real trouble. Luckily I had the wherewithal to recognize I needed help. I could not treat patients in this condition. I excused myself from the patient and went back to the lounge. They were laughing and having a grand time. Ted had a strange look on his face. They were all high. I knew I had to call in reinforcements. I called my senior resident and told him what I thought was going on. Well, it was quite a kerfuffle.

Nursing supervisors and hospital administrators descended on the area. The brownies were confiscated. The nurses and I were sent home. Two nursing supervisors manned the area until the morning shift replacements arrived. My senior resident treated the asthmatic patient and stayed the rest of my shift. Ted admitted he had spiked the brownies with marijuana. He truly meant no harm. He was trying to be funny and a part of that social buzz in the 4 am lull. He was part of the buzz alright, we all were. From one perspective it was rather funny – a whole late night ER staff and one earnest intern high on marijuana brownies. But there could have been serious consequences and Ted had clearly not reckoned with them. He thought he was being a cool, groovy guy. Fortunately, no one was harmed as best I know. Ted lost his job but was not reported to the criminal justice system. That was probably a good decision. His act was not malicious. All involved, me included, would have gotten tied up with the red tape of legal testimony. The hospital would have been seriously embarrassed.

As for me, I was left with a good story to tell and I gained a little "street cred" among my peers and the hospital hierarchy. I enjoyed the smiles and knowing looks I got over the next few days. But overall it was kept pretty quiet and everyone quickly moved on. At Presbyterian Hospital there was important work to be done. But the impression lingered for me. It was New York City and the Graveyard Shift. Things can get a little strange.

Dark and Low

It was one hard year. Looking back on it 37 years later I think it is the toughest sustained thing I have ever done. It was physically very demanding. Most rotations required spending every third night in the hospital with limited sleep; and the two ICU rotations were every other night in the hospital. So most work weeks were 120 hours or more; and we frequently worked 30-36 consecutive hours with little sleep. I did not object to that. I was young, healthy, and had anticipated the heavy work load. Modern day residency training programs have limitations on consecutive and total hours worked. The rationale behind these limitations is to reduce medical errors due to fatigue. There is some validity to that but I do not agree that modern training is better for the residents or the patients. Medical illnesses do not proceed by time schedules. The best way to learn the progression of an illness is to be there for most of it. The best way to know every detail about your patients is to be responsible for them for the great majority of the time. Handing off details and responsibility every 8 to 12 hours for a sick patient can miss details and cause errors. These can be compounded with each transfer of responsibility. It is also inefficient. Time spent transmitting the information to a new shift of residents could have been spent observing and caring for the patient. We groused and grumbled about the long hours but we understood the necessity of them. Using that 37 year perspective I can say with full confidence that the training I received was invaluable.

What I did not anticipate was the mental and emotional strain. There was enormous pressure at multiple levels from multiple sources. The Columbia Presbyterian philosophy was that interns learned medical decision making best by being fully responsible for those decisions. Yes, we had to be accountable for them and review them with more senior residents and attending doctors, but many times it was just us and the patient. We were it. This responsibility for decision making often caused me much anxiety, particularly in the first several months of the internship. All medical students are taught – first do no harm – primo non nocere. It is in the oath we take. So every time you made a decision you did not want to do anything that could harm the patient – but you also did not to want to omit something that could help the patient. What is the best decision for this patient at this moment of their illness? It was often a hard call for young, aspiring doctors.

Pressure to do well and make the right decisions came not just from concern for the patient. You wanted to be thought a good intern by your immediate superiors, the junior and senior medical residents. You wanted to be a "good doctor" in the eyes of the medical school professors and attendings. You wanted to live up to the honor of being chosen one of 16 interns to train at the Columbia Presbyterian Medical Center. And there was competition with those other interns – you wanted to do as well or better than they in becoming a good doctor. You wanted to be respected by the nurses as someone who knew what he was doing and not be thought a bumbling idiot. Lastly and importantly to me, I wanted to satisfy myself that I could handle it. Did I have the "right stuff"? We were all talented people, and not just academically. I suspect most of us had done well in most of the things we had tried to that point in our lives, so failure was not an option. But this internship year was so very hard.

There was so much to learn in addition to the medical facts of all the medical diseases. Yes, I know what diabetic ketoacidosis is, but exactly how do I treat it at 10 pm at night? How much insulin do I give? How often? How much IV fluid? How frequently do I check blood sugars, etc.? Do I call the senior resident or not? What do I do first, when I am responsible for 3 sick patients at 2 am? Do I believe a nurse when she calls at 4 am and says I need to see a patient right away or can it wait while I finish with something else?

So in mid-January of my internship year I got pretty down. I had been going full tilt for 6 months. The challenge and excitement of being a new doctor had worn thin. The sustained effects of sleep deprivation were probably causing a toll. During the months of November, December, and January we literally never saw the light of day. Winter daylight is short in New York City and we were always in the hospital. So sleep deprivation and lack of exposure to light were catching up to me. Looking back I think that I was probably depressed. I distinctly remember wondering to myself if I could make it to my 10 day vacation in mid-February. I badly wanted the year to be over. It was dark outside and my spirit was dark and low. In some weak moments I even entertained the thought of bailing out, quitting. I knew I never would, but I had the thoughts – and they startled me. I did not talk about it with anyone. Julie was 5 years into her career and very busy in her own sphere. Whenever I was home I was cranky and probably slept most of the time. My life was completely focused on the hospital and getting through each day. So we never really discussed how low I felt.

A little thing happened that seemed to help me get through this period. My dad called one day out of the blue, said he was in NYC on business and could we meet. I had not seen him since med school graduation 7 months before and had talked to him briefly just once or twice since then. I was able to slip out of the hospital and meet him for dinner in a local Washington Heights dive. We didn't talk about much, simple every day family things mostly. I must have given some indication of how down I was but I don't think he gave me a pep talk per se. He was just there, just visiting a son trying to become a doctor. For some reason my spirit seemed a little lighter after that visit. Over the next few days I knew I would get to my vacation in February and I knew I would make it through the year. So an unannounced visit from a dad gave a boost and a little needed toughness to a struggling intern. It's interesting that small shows of simple support can sometimes make unintended big differences. The dark and low internship blues had been lifted, not completely, but enough. I was ready for whatever came next.

The Upside Down Bedpan

Our 9 West rotation was an all-female ward service. I found it harder to care for the women patients. Their histories seemed to take longer to elicit. They often remembered more detail and past symptoms than the men we saw. Many of them had more chronic diseases with more past hospital admissions. And the physical exam took longer. Minutes count for interns. Anything that slowed us down lessened our efficiency and lengthened our days. One part of the exam that added time to the admission process was the pelvic examination. We had been taught that a complete physical exam was to be done on each hospital admission. For women that included a pelvic, the internal exam of the cervix, uterus and ovaries. Early in the year I had admitted and worked up a 70 year old lady with pneumonia. It was around midnight and the ward beds were all full. In this instance the nurses wheeled an extra bed into the hallway right outside the nursing station and surrounded it with mobile curtains on wheels. If I was to do a pelvic exam it would be in the hallway with little privacy. I skipped it to save time and give the patient a break. She was tired and sick. The last thing she wanted at that hour was a pelvic exam. And I did not think the pelvic exam was pertinent to her current problem of pneumonia.

On attending rounds the next morning I was asked why I thought the patient had pneumonia.

"Has anything predisposed her to an infection?

"She smoked a pack a day but quit 20 years ago."

"Any reason to think she is immunosuppressed?"

"Not really, she is not diabetic and has no history of cancer."

"I did not hear you describe any findings on exam other than the lungs. Any pelvic masses?"

Right then I knew I was in trouble. This lady could not remember if she ever had a pelvic exam. Gynecological cancers, particularly ovarian, can be silent before causing grave issues. A patient with a growing ovarian cancer can have a depressed immune system predisposing them to infections. Unlikely in this case, but I could not rule it out. I did the pelvic later that afternoon. It was normal but the attending had made his point. A complete exam on a woman with an illness causing hospitalization includes the pelvic exam.

From then on I took the time to do pelvic exams on all women admitted to the service. The first time I set up to do one the nurse handed me a metal bedpan. I thought she wanted me to ask if the patient needed to use it before the exam, certainly a reasonable question. But that was not the reason for the bedpan. Nope, the bedpan was turned upside down and the patient asked to rest her hips and butt on the cold metal pan so that her pelvis was slightly raised off the bed, making the exam easier on both of us. A portable, large, bright exam light was turned on and we did the pelvic. That is quite an image; midnight in a prestigious hospital, an intern, a nurse, a hospital bed in an open hallway, a bright exam light and the patient on an upside down bedpan.

I shudder now at some of the things we did. Those exams could have been done later in the hospital stay in a private exam room with an exam table equipped with stirrups. The patients would have been better served. That was one part of the Columbia Presbyterian philosophy that was a little misguided. But on 9 West we did complete exams on the patients prior to attending rounds and the dignity of some female patients was probably compromised.

In late March I had one more week on this service and then would be finished with it for the year. My next rotation was the day shift in Area B. This was

10 hour shift work, noon to 10 pm. I would get to sleep every night for 3 weeks. We all knew this was our easiest rotation. It was a cakewalk compared to the rest of the year. I was really looking forward to it.

But my resolve was tested again. One of my fellow interns, Adele Tedeschi, had come down with hepatitis and had to be out of work for 4 to 6 weeks. She was due to relieve me on the 9 West service. Someone would have to replace her on that rotation. That someone turned out to be me. Our Chief Resident, John Bilezekian, came up to me one day after attending rounds. He said the only thing to be done was for me to continue on 9 West for another 4 weeks until Adele was well enough to work. I would lose my Area B day shift rotation. That could be covered easily by a second year resident and a part time attending – an intern would not be missed. But an intern would surely be missed on 9 West. I stammered out several protests – but John, I am just finishing this rotation; I need a break from the wards; there must be another intern who could do it. It all was to no avail – he thought I was the one best equipped to handle it. The schedule would work for intern X but John thought I was more stable and could handle the extra stress and work load better. I was not flattered. I was steamed and feeling sorry for myself. I thought I could see the light at the end of this dreadful internship slog but that light had just dimmed. I would have been refreshed by the stint of day shift work in Area B and then it was just 2 more months until the end. Now I had to deal with 4 additional weeks of the grind of inpatient work on 9 West before those last two months. Yep, more sick little old ladies, and more late night pelvic exams on upside down bed pans.

Checked Out

There is a confidence that comes with completing an arduous journey. In mid-June I was feeling that confidence. I had been a doctor for almost a year and had learned more than I thought possible to learn in one year. In two weeks I would no longer be an intern. The work load as a resident would still be heavy but not nearly as bad – on call every fourth night rather than every third. And someone else, the new interns, would be doing all the scut work and be first on the firing line. There was still much to learn, but the learning would be different as a resident and easier to absorb.

There is one good day out of three when you are on call every third night. On the middle day of the cycle you would have slept at home the night before and you were not on call for any new hospital admissions. You felt fresh and less pressured. The work load was more defined and less open ended. On one such afternoon in June I was circulating on the 8 West service, checking on my 10 patients before signing out to a fellow intern on call for the night. It was a fine, sunny early summer day – the kind to lift even a medical intern's outlook. My routine was to stop by a patient's bed, check the vital sign sheet, talk with the patient a few minutes, do a brief, targeted exam and then write a progress note in their chart. I was making my own work rounds, checking on things at my own pace. I was thinking about the patients and making plans about their treatment for the next day or two. Who needed what test, which drug dose to change, and who was near hospital discharge? I felt in control and I felt like I knew what I was doing. I was a Columbia Presbyterian Medical Intern pretty much on the top of his game. In short, I was being a doctor.

I was also styling. We interns all developed our own mannerisms and personalities as new doctors. Some wore their short white lab coats with the pockets bulging with equipment and notebooks. We all wore white pants. My look as I was working and not on formal attending rounds did not include the white coat. I had on the white pants and a dress shirt with a tie, trying to be neat and professional. My sleeves were rolled up and folded just so. I had a stack of index cards in my shirt pocket – one for each of my current 10 patients. Each card had my own short hand notes on the pertinent details on each patient. I had it all in my head but I liked keeping those index cards as well. It was nice to stack them away when a patient was discharged. I carried my stethoscope in my back rear pocket, not draped around the neck as many docs do now. I had an elastic tourniquet tied around a belt loop. It had Velcro on it – easier to tie around a patient's arm to draw blood. I had my favorite pen in my shirt pocket for writing progress notes. My MO was to be efficient and in control, but to also look the part of the Cool Presbyterian Doc.

On this particular day I became aware of two
women looking around on 8 West as I moved from one
patient to another. They were the two hospital
Infectious Disease Control/Epidemiology nurses. They
worked for the ID service and helped the service and the
hospital keep track of infections. They defined
procedures to limit hospital acquired and transmitted
infections. Part of their job was to make sure we interns
washed our hands and used proper sterile techniques in
taking cultures. They were both blond, attractive
women. They had also been to the beach. Even a worn
down intern with his nose to the grindstone couldn't
help but notice two attractive, blond women in tight
nursing uniforms sporting new suntans. I knew who
they were, but did not think they really knew me. They
usually got their information from the residents on the
service or the nurses on the floor. We interns were
usually too busy, too grungy, and too grumpy to talk to.
And we wanted to avoid being criticized for some
breach in infection control technique. Well, on this
particular day they seemed to want to talk to me. They
were organizing a study on patients with tuberculosis.
Did I have such a patient? No, I did not. Would I
please contact them if I got a new one admitted to me in
the next few days? Well, I sure would. Their smiles
and flashing eyes left little doubt of that. And isn't this
nice – attractive nurses were interested in what I had to
say. They left and I got on with my work. I felt even
better as I finished up for the day. I was pretty sure I
had been checked out by two attractive and subtly

flirtatious nurses. This business about being a doctor just might have its upside.

Part 2: The Residency Years

No Longer an Intern

Be careful what you wish for. We interns had all yearned for and pointed to the day we became first year residents. We couldn't wait to leave those internship blues and responsibilities behind. As my man, Jimmy C, had said through the fog of his fatigue, I too, could not wait to hand that ball off to someone, anyone, else.

So on June 30, 1975 I finished my year as a medical intern. Then at 8 am on July 1, 1975 I became the first year medical resident (PGY2 in today's terms) on the 8 East/9 East Medical Service. There were 3 fresh faced, apprehensive, new interns looking at me for their first day's marching orders. Suddenly, with one turn of the calendar there were new expectations and new responsibilities. I realized that instead of just myself and my 10 patients of yesterday, I was now in charge of these 3 greenhorns and 30 sick patients. There was nothing to do but get started.

As first year residents we were on call in the hospital every fourth rather than every third night. With different responsibilities and less scut work to do, rarely were we up all night and often got a few hours of sleep in the hospital. That made a world of difference in your outlook – you almost felt like a normal human being again. You also had more power. There was someone below you on the food chain.

But you were expected to guide and teach those new interns. The Latin for doctor is docere – to teach. I remembered being grateful as an intern when a resident was helpful and encouraging rather than disdainful of my ignorance. You realized you did have a lot of knowledge to impart. But there was still much to learn. The attendings expected you to read more, be more aware of the details of diseases and the intricacies of differential diagnoses. The pressure was different but it was still there. I had come a far piece down the road but I was not yet complete as a doctor of medicine.

The Dangers of the DTs

The first few weeks as a resident seemed fairly smooth. I had begun to get a handle on most of the 30 patients and on the different personalities of the 3 new interns under my wing. There had been no disasters and no obvious disapproval from the attendings on the service. One of my interns was a Stanford grad, quite bright and a quick study. He seemed to have stability and the "right stuff" to get through the rigors in front of him. One afternoon about 3 weeks into our service together he paged me with a note of panic in his voice. I had been with another intern up on 9 East.

"Jeff, you gotta get down here! Mr. Y has gone nuts. We have real trouble."

Mr. Y had been admitted 3 days prior with an empyema, a collection of infected fluid between the lung and his chest wall. He had been really sick with high fever. He was still bedridden on rounds that morning. He had chest tubes in place to drain the foul smelling infected fluid. As I bounded down the stairway and into the 8 East ward I figured he must have gone into septic shock.

To my utter amazement Mr. Y was up out of bed. There was broken glass and foul smelling empyema fluid all over the floor. In his hand he held the jagged bottom half of the glass bottle into which his empyema fluid had been draining. He was waving that dangerous shard of glass at the world in general. Standing between him and the hallway, where he seemed intent on going, was the brave but rattled intern, Dr. R. Mr. Y was agitated and raving about being in prison and tormented by his doctors.

I edged up to Dr. R. "IV valium, 5 mg now!"

"Can't. He's pulled his IV half way out."

I wasn't sure anyone could get close enough to give him an intramuscular injection without being hurt by that jagged glass shard he kept pointing at Dr. R. I asked the nurses to call security, but I was not sure they could handle this situation. We would have to talk him down. And we did. Both of us spoke in gentle tones reminding him that he was sick with an infection, was in a hospital, and that we were doctors trying to help him. We were helped by the fact that under all that paranoia he was still a very sick man. This explosion of energy to get out of bed, grab his chest tube bottle, smash it on the

floor and brandish it as a weapon had drained him. He
let the remains of the broken bottle drop to the floor.
We guided him back to his bed, and gave him a healthy
dose of an intramuscular sedative. Whew, that was
close.

Security personnel and nursing supervisors had arrived
but the drama was pretty much over. There was a mess
to be cleaned up and his chest tube would have to be
reconnected to a new set up. Luckily he had not pulled
the tube completely out of his chest.

Dr. R and I stepped away to review what had happened
and where to go next.

"Did you have him on Librium? What set him off?
Have you paged the chest surgeon?"

I had fired questions in rapid sequence, still not quite
believing what I had just seen.

Delirium Tremens is the term. It is the agitated,
delirious, sometimes violent state that can occur after
withdrawal from heavy use of alcohol. Bad episodes
like this one are infrequent and usually occur only with
real hard core drinkers who have consumed large
amounts daily for months on end. The time frame is
usually 3 to 4 days from the last drink. We knew Mr. Y
was an alcoholic – that is one reason he had such a
serious infection – he was malnourished and
immunosuppressed. But we did not know he had any
history of prior withdrawal issues. The norm back then
to cover potential alcohol withdrawal in hospitalized

patients was to give them standard doses of benzodiazepines. Dr. R had done so with Mr. Y, but the dose had been the usual moderate amount for a patient not known to have a history of the DTs. Mr. Y had been lethargic and bed ridden since his admission due to the severe infection. If we had over sedated him we might have compromised his respiratory function which was already impaired from the pneumonia and empyema. His improvement from the chest tube drainage and the IV antibiotics over 3 days coincided perfectly with his 3 day withdrawal from heavy alcohol use. A perfect storm of paranoia, delusion and improved strength resulted in the weird, potentially violent scene that day on 8 East.

Mr. Y had his chest tube drainage bottles replaced and his tubing resecured by the chest surgical resident who was not happy to be called back for this chore. He listened to our explanation, but his look clearly indicated that we somehow must have screwed up. Mr. Y completed another 3 weeks in the hospital with slow improvement. The healthy doses of sedatives were slowly reduced and he was discharged without further issues. In our more modern era chest tube drainage bottles are no longer glass. They are fiberglass/plastic composites that are probably indestructible – and certainly much harder to weaponize in a deluded alcohol withdrawal state.

Tuesday Noon Conference

Conferences and formal discussions are part of the scene in academic medical centers. Interesting cases and related topics are reviewed. The medical house staff had a conference each Tuesday at noon. All interns and residents were expected to attend – but interns were usually too busy or too tired to attend most. I may have made a third of them as an intern and did not absorb much due to fatigue. Topics were discussed by fellow residents or by young attendings. Often a more senior medical school professor would drop in if the topic was of interest to him or her. The idea was to review current medical literature on topics pertinent to some of the patients we saw. We discussed the underlying pathophysiology of disease, how it manifested in the patients we saw, and what was the current thinking on how best to treat it. We were trained to critically review the data in the literature, compare it to what was already known, critique it for its thoroughness and statistical accuracy and see how it might apply to patients. Professors constantly admonished us not to accept all published data on face value. We were to think critically and carefully about new and old data. Did it really hold up? Aspiring doctors have to make critical decisions not just about patients in the clinical setting but also about all the new medical information and published data that comes along. In retrospect the most valuable part of the Columbia Presbyterian training was the constant emphasis on how to think critically. Weigh all the data, piece it together, and then make the decision that makes the most sense.

The Chief Medical Resident (usually a PGY4) organized the conferences. As a junior (first year, PGY2) resident we were asked to research a topic and speak on it for 20 or 30 minutes and then take questions. We were assigned to do this 2 or 3 times during the year. I happened to have kept my notes and preparation for 2 of those conferences from 37 years ago. I recently reviewed them and was astounded at the detail of anatomy, pathophysiology, literature review, and clinical manifestations in my presentations. Both topics are far from my area of practice (cardiology) after my medical training. I seemed to know far more then than I do now about the intricacies of Graves' disease and Membranous Glomerulonephritis. I wonder now why I chose those topics. They had to have been related to a patient on the service in the recent weeks. If the discussions in those conferences were at the level that my old notes seem to indicate, we were a pretty erudite group of young doctors.

The Harkness Pavilion

Edward S. Harkness was an American philanthropist who died in 1940. The pavilion in his name and funded by his estate is a ten story tower connected to the main hospital building by doors on the fourth and ninth floors. Private patients with their own physicians were hospitalized in single rooms in this section of the Medical Center. As interns we did not have any assignment in this area of the Columbia Presbyterian Medical Complex. But as first year residents we crossed those doors from the medical ward services into the realm of private patient care. It sure was different. In general the patients were not as "sick". Their illnesses ran the gamut of medicine but sometimes it was hard to figure out just why their docs had hospitalized them. Many patients were from the upper crust of NYC society. In that era of the 1970s it was not uncommon for such a patient to be hospitalized for a little "R and R", to escape a difficult, stressful situation in their lives. I recall a patient or two staying a week getting vitamin injections every other day. This kind of medical care does not exist today, and properly so.

It was quite an adjustment coming from the ward services where we made all the decisions and wrote all the orders. On our Harkness rotations as residents each patient had his/her own private doc who called all the shots. Many were good and made an effort to engage and teach us. But some just wanted us to do what they told us to do and be there at night if there were problems. There were no formal teaching rounds in the morning. We were on our own for learning about the diseases we saw in those patients. Our nickname for the Harkness Pavilion rotation was "The Darkness".

To brighten things, every once in a while the word would go out among the medical residents. "Guess who I admitted to the Darkness last night." More than a few of the celebrities of the day entered the

Presbyterian Harkness Pavilion for various medical issues.

Cancer chemotherapy was in its youth back then. Things were done that would be unthinkable now. One of our duties on the "Darkness" rotation was to give intravenous doses of certain agents to some cancer patients. One of the agents used was Adriamycin. We called it the Red Death. It came in small vials as a red liquid. Depending on the dose ordered by the attending you had to break off the top of 10 to 20 small glass vials and draw up the right dose in a huge syringe. You were then supposed to inject it intravenously over 10 minutes. We had not been trained to do this. I hated it. Looking back I cannot believe we were required to do that on our own, and often in the middle of the night. It was dangerous on many levels. The stuff was caustic. If the IV infiltrated or leaked the surrounding tissue would be damaged. If any leaked while you broke open the vials it could burn your skin. You could easily cut yourself on the glass as you broke open the vials. You could miscalculate and give the wrong dose of that red poison. But there was no one else, no trained pharmacists to do the precise, sterile preparations getting the drug into soft plastic infusion bags; and no special chemotherapy nurses to infuse it slowly and carefully over a precise time period. It was just a Columbia Presbyterian medical resident trying to get through his rotation in the "Darkness" Pavilion.

One of the best stories about a resident's Harkness experience came from Dr. Mike S who was one year ahead of me. One morning he spun out his tale to a few of his colleagues as follows: He was at the 9th floor nursing station at 2 am speaking to the floor nurse. Many Harkness patients had "private duty" nurses in addition to the regular hospital floor nurses. These were generally older, retired nurses paid to sit in a patient's room and respond to their every need. As he looked down the darkened hall he saw an old white haired lady slowly crawling along in a flimsy hospital gown. She was on her stomach using just her elbows to move along and was calling out for help. He and the nurse ran to her. It was an 80 year old lady previously bedridden with a stroke. They knew she had a private duty nurse and could not figure out what she was doing crawling along in the hallway. As it turned out it was not the patient who needed help. Her 65 year old private duty nurse had collapsed at the foot of her bed. The patient slowly and painfully had crawled out of her bed and into the hall to get help. Dr. S ran into the room. The private duty nurse was dead on the floor. So not only did he have to calm a freaked out patient stomach crawling along the hallway floor, he had to pronounce a private duty nurse dead and start the bureaucratic machinations required in the death of a hospital worker. It was just another night in the Darkness Pavilion.

Glenda Rounds

Some people are born to teach. Glenda Garvey was one of them. In 1975 she was just a year out of her fellowship training in Infectious Diseases. That year she became the Medical Director of the medical ICU. As junior residents we had two separate 4 week rotations in the medical ICU. That meant a lot of time spent with Dr Garvey.

As interns we had a huge workload and a lot of responsibility in the ICU. As I indicated earlier, the complexity of the ICU patients was high and although expected to know everything about the patients, interns were not expected to make all the complex management/treatment decisions. That was the responsibility of the junior resident on the service. The responsibility was intense but as residents we felt we learned an awful lot about serious medical illnesses due to that responsibility. The ICU patient mix in that era included people with diabetic ketoacidosis, pulmonary edema, septic shock, GI bleeding, hypovolemic shock, respiratory failure, and drug overdoses. True to the Presbyterian Hospital philosophy we learned to make decisions by being responsible for and making them on the fly in the clinical situation. Those decisions were reviewed and discussed the following morning on attending rounds, as they were on the medical ward services. But these formal attending rounds in the ICU were secondary to our time with Dr. Garvey. We called them "Glenda Rounds" and that is where most of our learning and critical thinking occurred. The process was an informal give and take. Dr. Garvey and the resident would go from patient to patient from 9 am until attending rounds at 10 am reviewing everything and I mean everything. We would often continue for another hour after attending rounds and sometimes recommence around 5 pm before evening sign out. They were true work rounds since we responded to problems and adjusted patient treatment as we went along.

ICU care was a fairly new process in 1975 and Dr. Garvey was learning, just as we were. We all called her Glenda.   We learned ICU medicine by seeing it and doing it, making the decisions. Glenda was right there with us, making us be careful and think critically, but also supporting us.   She was the prototypical young medical school professor. She could be funny and irreverent, and at times seemed more like one of us than the faculty person she was. Her persona was elegant, erudite and warm. Her style tended to long skirts and white blouses. She wore horn rim glasses and her long dark hair pulled back into a bun.   It was a stylish, attractive package. More than one male resident had a crush on her.

But Glenda was also very compulsive. Her rounds at times seemed to drag on forever. She needed all the details and had us residents double check all the treatment options.

"Glenda, we already went over that."

"Yes, but I just thought of something else."

And what she had to offer always seemed to make sense. Glenda Rounds taught us all so very well.

It is a common phenomenon to look back on one's education and training and pinpoint a certain teacher or professor who really made a difference in your life and work. There were a few at CPMC, but for this old doc, and I am sure many others, Dr. Glenda Garvey stands out above all the rest.

She was a graduate of Columbia University Physicians and Surgeons Medical School and did her residency and fellowship training at CPMC. She was Columbia Medical School and Presbyterian Hospital to the core. She was first appointed as faculty in 1974. 1975, my year as a junior resident, was her first as the ICU director. She went on to direct the ICU for 20 years. During that time she instructed every medical house officer who passed through the program.

She also directed the third year medical student clerkship program for 20 years. She went on to receive numerous academic and teaching awards (teacher of the year multiple times, distinguished professor awards). She died in 2004 at the age of 61, a huge loss to the medical school and the residency program. There is now a Glenda Garvey Teaching Academy at the medical school and hospital - in her honor - and to propagate her excellence and teaching ideals. None of this is any surprise to me. There could be no finer role model for young, aspiring doctors. I am proud to say I knew her back when.

TGIF

The internship year for me was a total grind. I got through the long days as best I could and went home to sleep whenever I had finished. The time in the hospital was all work, responsibility, and anxiety. As a resident things were a little different. There did seem to be some time for relaxation and a little fun; not much, but just enough to make me think that doctors could be human beings after all. After a month or two as a resident I became aware of TGIF. It was probably Peggy Fracaro, one of those attractive Infectious Disease nurses who clued me in.

TGIF was a weekly gathering around 5 pm on Fridays in the lounge area on the second floor. The second floor of the hospital was the house staff sleeping and locker room quarters. There were rooms, beds, and shower areas for the interns/residents who were on call on a particular night. The attendees were mainly first and second year residents plus a few nurses. It was very rare for any intern to have the time or energy to make an appearance. I was not even aware that this social rite existed during my internship year. We would relax with some snacks and a glass of wine or two. Stories would be shared, gripes aired, and rumors swapped. Particularly interesting or unusual patients might be discussed as well. Occasionally a young faculty attending like Dr. Garvey or Dr. Neu would drop in. I began to realize that the medical house staff had a social nerve center where a lot transpired. We were able to unwind and have some laughs. This simple social gathering served dual functions. Yes, it was the social nerve center of the house staff but it was also a healthy outlet where some of the heavy tensions and responsibilities of our lives could be viewed with some humor. I became a regular.

The Box

All hospitals have a cardiac arrest or "Code" team. When a patient is found unresponsive or in extremis, a "code" is paged. The code team consists of an anesthesia resident, a surgical resident, and a medical resident. Each has a specific job or role in a cardiac arrest. The medical resident is effectively "in charge". He/she calls the shots, the meds to be given, the shocks to be delivered. So as a medical resident on call in the ICU or on the wards, our duty was to respond to and direct any cardiac arrests that occurred anywhere in the hospital. In the mid 1970s we had defibrillators, but they were crude heavy beasts. The days of lightweight, sleek, portable AEDs and small implantable defibrillators were still in the distant future. Our devices were large, cumbersome and heavy compared to today's portable ones. The CCU and the medical ICU had devices which were on carts. Statistically, more arrests occurred in these locales. There was also one in the ER. But the rest of the hospital did not have defibrillators. If an arrest occurred on the medical, surgical, orthopedic, or gynecologic ward services, or in the Harkness Pavilion, the arrest team would respond to the page and rush to the location. The portable defibrillator, The Box, arrived with the medical resident. It had to be carried to wherever the arrest had occurred. This was no mean feat. The device was roughly 18 by 18 by 8 inches. It rested on the floor and had two plastic handles extending a foot above its top surface. It was also heavy. I do not recall its exact weight but it had to be in the range of 20 pounds. When an arrest was paged you had to assess where you were vis-a-vis the location of the arrest; then plan your sprint to the site. Up 3-4 floors, you carried the box. Up any more than that you might have to chance the time it took to take the elevator. Going down several floors was less of an issue. But going down a floor or two could have its problems. One of my fellow residents stumbled going down a flight of stairs with the Box. She fractured her kneecap. She was one tough cookie and did not miss a day of work. But she did not carry the Box for quite some time.

The real problems came on the occasional arrests in the Neurological Institute or the Eye Institute. These were separate towers from the main Presbyterian Hospital. You had to take underground tunnels for a block or two, then sprint up stairs or find the elevator. I am not sure that any patient needing defibrillation in those outlying spots ever survived. It took too many minutes to get there with the Box. One could argue that the occasional resident sprinting the equivalent of two to three blocks and several flights of stairs lugging a 20 pound Box might need resuscitation as much as the poor patient.

Club 22

Like TGIF there was another social gathering that had eluded my awareness during my internship year. This one was irregular and weather dependent. I did not learn of it until a fine early fall day in my first year residency. Attending rounds had finished on the ward service where I was the first year resident. There was a lull in clinical and academic activity. It was around lunch time so I made my way down the hall toward the small house staff library/lounge on 8 Stem where I kept my bagged lunch.

"Hey, Dr. Blake. Ever been up to Club 22?" Peggy Fracaro, the hospital nurse epidemiologist, had sidled up to me as I strolled down the hall. She had become friendly with many of the residents in my year, me included. She was a fixture in the social ebb and flow of the medical house staff. She always seemed to have her finger on the pulse of what was happening in the whole hospital, not just on the medical or ID services. Her job required that she circulate throughout all the services. But she seemed to like hanging with the medical house staff best. We all liked her and why not. She was smart, pretty, and a source of all kinds of hospital gossip.

I had no idea what she was talking about. Club 22? Was she referring to some hot night spot in the city? She saw my puzzlement. "It's the roof, silly. You should check it out. Come on up and have lunch with us."

The elevator of the hospital did in fact go to the 22nd floor. I was amazed as I got off onto the flat gravel rooftop of the Presbyterian Hospital tower. There were several nooks and crannies plus pieces of mechanical equipment. The wall of the roof was bordered by a chest high concrete wall. The day was sunny, warm and clear. You could stand at that wall and have unobstructed views in all directions. To the south we could see all the way to the tip of the island, and since the day was clear without smog, Lady Liberty was visible out in the harbor, a full 10 miles away from our perch at 168th street in upper Manhattan.

A few other residents and nurses were sitting in outdoor lounge chairs enjoying the sun and their lunch. Adele Tedeschi and Rich Mattern were among them. They were husband and wife, a tough thing to do as house staff, but they made it work. A beeper went off. Adele had been paged. I assumed she would head for the elevator but she calmly walked over to a wall phone next to the elevator to answer her page. Not only did this "club" have spectacular views but it came equipped with beach chairs and a phone. You could spend a little time up here and not be missed on the wards if you were careful. How had I missed this? How come Adele and Rich knew about it before I did? I was not sure what the attendings and various hospital powers thought about this little retreat, but it became a spot I enjoyed when the weather was good and I had the time for a quick outdoor lunch. Somehow the word would spread. "It's a Club 22 day. Don't' miss it." Like TGIF, Club 22 was a scene for socializing and some release of the grinding tension of our work. The opportunities were limited and the work was tough. We grabbed them when we could.

The Power of the MAR

As it did on July 1, 1975 my role and responsibilities again changed dramatically on July 1, 1976. I became a senior (second year, or PGY3) resident. Senior residents were at the top of the 3 year pyramid of the medical house staff. We had more time for elective rotations on the specialty services like Renal, GI, Endocrine or Cardiology. We were expected to be more scholarly and read more of the medical literature and textbooks. We could spend up to two months doing research in a clinical specialty if we chose. We were supposedly solidifying all that clinical know how we gained in our first 2 years – making us more complete as doctors and internists.

Our on call clinical responsibility for the medical ward services occurred every 5th or 6th day. During that 24 hour period we were the Supreme Being for the medical wards. The most vexing or difficult problems were brought to us for consultation if no attending staff were around. More importantly we were the Medical Admitting Resident, the MAR for that day/night. Our role was to evaluate patients in the ER, the Area B walk in clinic, or other areas of the hospital (the Neurological Institute, the Surgical Service, etc) that another intern, resident or even attending felt might need admission to or transfer to the Medical Ward Service. We had complete power over those decisions. With that complete power came complete responsibility. If we saw a patient in Area B and decided they were not sick enough to merit admission they were sent home. It was a fine line to walk. Not admitting someone who was potentially seriously ill was an error no one wanted to make. Yet beds on the ward services were often limited and scuttlebutt among interns and residents on the ward services had a way of filtering down to us. As an MAR you did not want to have the reputation of admitting every borderline patient who might need hospitalization but really was not that sick. The interns and junior residents knew all the senior residents as either "Rocks" or "Sieves" when they were on call as the MAR.

As an intern signed out to a colleague for the night he might ask, "Who's the MAR?" and then wince with sympathy if someone known to be a "Sieve" was on call. It could mean a very difficult night. I clearly remembered being an intern and all the work required of getting a new admission at midnight. If we got more than 4 admissions in a 24 hour period we hated the MAR, whether all those admissions were justified or not. So as MARs most of us were more comfortable coming down on the Rock side as opposed to the Sieve side.

Many of the admission decisions were fairly straight forward. Patients with an ICU type illness (GI bleeding, shock, MI, pulmonary edema, ketoacidosis, etc.) were quick and dirty; a brief look and review of the data, then the call upstairs to the receiving intern. The

tougher calls were those patients with pneumonia who
did not appear too sick or the patient with moderate
heart failure who could get a dose of IV diuretics in the
ER and then go home in a few hours. Will they do all
right as an outpatient or would they be better in the
hospital? That decision clearly revolved around the
clinical data on the patient but it was also a function of
bed space availability and our comfort with the work
load we placed on our junior colleagues up on the ward
services. I think we learned to make quick solid clinical
decisions. But it was a different era. You had to be
pretty sick to get admitted. That was our house staff
culture. In retrospect, and certainly in today's terms, I
suspect that I sent more than one patient home who
might have benefited from a hospital stay.

MARs did have one compromise possibility. We
could bail out and admit a patient to the "Overnight
Ward". This was a small 6 bedded ward service across
the hall from Area B. It had a skeleton nursing staff and
was basically a holding area for the ER and the walk in
clinic. We could "admit" a patient there for 24-48
hours and care for them ourselves. We were the intern,
resident, and attending all rolled into one. There were
no rounds and no formal review of patient care in that
area. Senior residents were trusted to be clinically
sound in their use of this area. You could put a patient
there with mild to moderate heart failure and give them
IV diuretics for 12 to 24 hours, then send them on their
way. Patients with sickle cell anemia and moderate
sickle cell crises were often sent there for 12 – 24 hours
of IV narcotic treatment. Asthmatic patients in need of a

12-24 "buff up" were frequent visitors. All MARs
would use the Overnight Ward, particularly on nights we
had already admitted a large number of patients to the
wards upstairs. It was kind of a limbo area for patients,
so I tried to use it infrequently. The champion overnight
ward MAR in my year was Dr. Joe Tenenbaum. He was
probably the most compulsive and compassionate doc
among my peer group. We all worked long hours but
Joe always seemed to go the extra mile and extra hour.
His patients loved him. As the MAR it was hard for him
to send any patient home. He always seemed to have a
few (or several) patients in the overnight ward
throughout the year. And more than a few of those
patients stayed for more that 24 hours. I think he may
have started his private practice from that cadre of
patients after he finished his training!

The Jumper

The responsibilities of a night as the MAR
usually involved a series of difficult decisions for
several hours broken by a few short 15 to 20 minute
lulls. During one such lull I wandered over to the first
floor cafeteria around 10 pm for a quick snack. The
medical house resident (PGY2) was nursing a coffee,
enjoying a similar lull from his duties upstairs. We were
chatting amiably, a mix of business and hospital gossip,
when we were startled by a strange crunching sound. It
was a warm October night and we were sitting near an
open window to the courtyard outside the cafeteria.

There was something about that strange mixture of thud and crunch. The mind somehow knows when an unusual set of signals comes in. Our eyes locked. Could that have been something or someone hitting the courtyard pavement from the wards above? It was dark out there, we could not see anything. We sat back down, and briefly debated what it could have been. Then my beeper went off.

"Dr Blake - 6758. STAT." That was the hospital administrator. Right then we both knew. Someone had jumped or fallen onto that courtyard concrete. We both rushed out the door, Dr. B heading to the stairwell to dash upstairs to the medical wards where his responsibilities were. I headed to the courtyard. An ER nurse, the hospital administrator, and a security man had just arrived. As I closed in it was clear there was a crumpled human being on the pavement. It was a difficult picture to digest. Bones and limbs at distorted angles; blood and tissue on the pavement. The angle of the neck said it all. There was nothing to do but check the absence of pulse and respirations. I noted the hospital gown, took a brief look at the distorted face, noted the time and pronounced the patient officially dead. Hospital administrative personnel would handle the rest. I looked at the administrator.

"Anybody know what happened?"

"We think he jumped from the 12th or 14th floor."

Floors 12 and 14 were the surgical ward services. Both wards have windows on the east side that open onto the courtyard far below. The windows are chest high and open easily. As I have noted, back in the mid 1970s the hospital towers were not air conditioned. The only venting and cooling options were those windows.

A surgical ward patient made his way out of one of those windows to his death that warm October night. The later scuttle butt had it that he had a history of alcoholism. The DTs may have struck again.

The medical ward services on 8 and 9 West also opened onto that courtyard. One of our medical patients could have been the jumper as well. Dr. B and I both knew that as we rushed out of the cafeteria a few minutes back. We both knew we would have felt in some way responsible had it been one of "our" patients. After dealing with the paper work of the dead patient, I made my way up to 9 West to find Dr. B and report to him what I knew. After I recounted what I had found and what I knew about the patient jumping from the 14th floor he looked at me sheepishly and said.

"You know Jeff, as I ran up stairs I kept saying to myself, "I hope it's Rose, I hope it's Rose."

By his look and tone I knew he was at least in part being serious. I was a little taken aback, but I understood where he was coming from. We fought all the time to keep sick patients alive. That was our mission. The opposite was failure. But sometimes the simple fact was that a certain patient dying could relieve us of a huge burden. How should we feel about that? How do we

reconcile that feeling? Should we even admit to
ourselves that we have that sentiment?

Rose was a patient with sickle cell anemia. She was in
her mid twenties and an attractive young woman. She
was a very difficult, manipulative patient. All the
medical house staff knew her from her frequent visits to
the ER, Area B, and her various hospital admissions
with sickle cell crises. Because of their frequent painful
episodes of vascular ischemia, some sickle cell patients
become addicted to narcotics. Rose was one of those
patients. But she also had clear, unequivocal evidence
of organ damage from her crises. You never could tell
if she was truly in crisis or just wanting a stronger dose
of narcotics just for the high of it. Rose made these
difficult assessments harder with her seductive and
manipulative manner. Sometimes you just knew you
were being had.

Dr. B had admitted Rose as an intern 6 months
ago and since then he had followed her as an outpatient
in his group clinic. He had tried vainly to understand
and help her issue with the narcotics, but to no one's
surprise had gotten nowhere. The knowledge that you
are responsible for a patient, try very hard to do the right
thing, but get rebuffed and used in the process can be
quite distressing. On the night of the jumper, Rose
happened to be up on 9 West. She had been admitted
the day before with another sickle cell vascular crisis.
Despite what he said, I suspect Dr. B was relieved it was
not Rose who had jumped. But I did admire his
honesty in voicing his hope that it might have been her.

We never talked of the episode again. That night, and in the coming days, other patients and other responsibilities intervened. But the episode brought to the surface a very complex intersection of our emotions and responsibilities as young residents trying to become good doctors. In that era we got no training on how to handle this. There was no forum to discuss such things. In retrospect I think we both were glad that it came up in the semi-serious, semi-humorous way it did. We were able to acknowledge, however briefly and incompletely, that such thoughts and emotions were normal human reactions. Our job as young doctors was to recognize that, overcome it and move on.

Windows on the World

Other than those days and nights as the MAR, the senior residency year at CPMC moved at a more studied pace than the internship and junior residency years. We had acquired buckets full of clinical and practical knowledge. In our third year we refined that knowledge into a more academic appreciation and grasp of internal medicine. There was more time to read about disease, discuss patients and attend conferences. We also honed our skills as teachers to our more junior peers. We just might be scholars and attendings in waiting.

The national board examinations in internal medicine loomed at the end of our third year. If you completed 3 years of residency in internal medicine and passed this examination you called yourself a Board Certified Doctor of Internal Medicine. It was a prestigious accomplishment and made one's career path smoother. Joining a practice or taking a teaching position could depend on your board certification. You also got a nice certificate to hang on your wall.

All self respecting residents completing the CPMC program were expected to take and pass the exam. Much of the exam involved clinical cases and clinical decision making. Historical, exam and lab findings on patients were presented. You were then asked questions about management, diagnosis and treatment. We all felt comfortable about that aspect of the exam because of our strong clinical training. But the exam also probed your knowledge of the breadth and details of medical diseases. There was a lot of esoterica that we could be asked about. So for much of our third year my colleagues and I spent time reading a textbook of internal medicine. It was a huge tome – Harrison's Principles of Internal Medicine. I still have my 37 year old copy. Most of my colleagues read a lot more of it than I had. I just hoped I had read enough to get by most of the detailed questions. I felt my clinical knowledge and management skills would hold me in good stead on the rest of the exam.

The exam took two full days and was given in scattered locations around the country. Because of the number of training programs in the New York area there were two exam sites in the city. A few of my colleagues and I traveled by subway to the NYU site in southern Manhattan. I found the first day to be quite a challenge, although I thought I did fine. My good friend, Adele Tedeschi, and I were discussing the multiple clinical case vignettes as we rode the subway home. I had been stumped by one of them. She said,

"Oh yeah, the patient with Wilson's Disease. He had Kayser Fleisher rings."

My spirits sank. Wilson's Disease?? Oh no! I had based all of my answers and management of this case on a completely different diagnosis.

Wilson's Disease is an uncommon, genetic disorder of copper metabolism. Copper accumulates in the body and leads to hepatic, renal and neurological impairment. The copper sometimes accumulates in the cornea of the eyes leaving characteristic brownish rings. I had missed those K-F rings described in the physical exam. Adele was smart and I knew she probably had it right. So I had blown a good 20 per cent of the clinical part of the exam; and that was supposed to be my strong suit. Passing the boards was not a foregone conclusion. People did fail. The national pass rate in that era was around 75 per cent. I felt some dread that I might fall into that ignominious group of failures.

The second day was better. I do not think I made any major errors. Adele, two other friends, and I felt well enough about our performance that we decided to go out for dinner. We were downtown so the decision was to hit the Windows on the World at the top of the World Trade Center. We had a fine dinner, a lot of laughs, and enjoyed the majestic view. Despite my pangs of uncertainty about my screw up the first day, we all passed the exam with flying colors. Three tough years and a difficult examination were behind us. The world and our future in it looked bright. But as we all know, those twin towers no longer stand. That fact and all its associated loss make the memory of that post exam dinner celebration a bit dearer.

Part 3: Chief Resident

One More Year

Columbia Presbyterian Medical Center and I were not done with each other. Early in my senior residency year I got a call from the secretary to the Chief of Medicine.

"Dr. Blake, Dr. Aranow wants to see you. Can you come over to his office?"

Henry Aranow was the acting chief of the Department of Medicine at the Medical Center. He had assumed the role last year on the death of Dr. Charles Reagan who had been the Chief for several years. A summons to see the Chief could mean many things, some not good. I wracked my brain searching for something, somewhere I may have screwed up. Did he know about Club 22? Was he upset about it? Did he want me to present a case or give a conference? Did he want to talk about a patient? Does he want to talk about my future aspirations?

It turned out he did want to talk about my future. He wanted me to be the next Chief Medical Resident. I was surprised and did not know what to think. It was a prestigious position and offered to residents who had done well. It was unspoken and unspecified, but doing that job for a year meant you had a good chance of landing an assistant attending faculty role at the hospital in the future. It was considered a feather in your cap. But it was also an administrative and somewhat thankless job. You had to organize, schedule, and generally oversee the performance of the entire medical house staff. When they had problems or got sick, they came to you. If one was too sick to work, you had to figure out how to cover those responsibilities. Complaints about house staff performance or their attitude also came to you. All of these people were smart and often very opinionated. The job was often a combination of putting out fires and herding cats. I had decided to pursue a cardiology fellowship after my third year at CPMC. Taking the chief residency position would mean postponing that career path for a year.

Henry Aranow was the quintessential CPMC attending professor. He was tall, white haired, and patrician in manner and appearance. His background was in endocrinology. He knew he was in an interim role. There was an active search for a new, younger, more vibrant Chief to revive the department and the program. But like all CPMC professors he took his position with great seriousness. He had a certain gravitas and I respected him. He was not a Mafia don but in many ways his offer was too good to refuse. On July 1, 1977, three years after starting as a green intern, I became the Chief Medical Resident at the CPMC.

Hot Town, Summer in the City

New York City was a tough, gritty place in the summer of 1977. A brutal heat wave had gripped the city in July. City finances were in turmoil. Crime rates were much higher than today. The Son of Sam serial killer was stalking the streets. The Yankees were in the midst of the Bronx Zoo, Billy Martin years. Things were a little crazy. Then, just two weeks into my term as Chief Medical Resident, the lights went out. On July 13, 1977 at 8:30 pm a lightning strike at a Con Edison facility in Westchester started a cascade of lost power. By 9:30 pm the entire city was down. It was hot and dark. No lights, no subways, no elevators, no air conditioning, no traffic lights. That is quite a combo for 15 million people. Looting, vandalism and arson occurred in many areas, including the Washington Heights neighborhood surrounding Presbyterian Hospital.

Hospitals did not have full back up generators as they do today. So in most hospitals there were no lights, no elevators, no monitors, no paging system, and no power in the ORs. Presbyterian Hospital did have limited generator capability. The ICUs and the ER did have some power, but that was it. Everything else was dark. This was not in the Chief Resident handbook. Julie and I were home in Old Tappan, New Jersey when the power went out. New Jersey was not affected. We had power. Old Tappan is a 15 minute commute over the GW Bridge to the medical center. We had moved out there the year before during my senior residency. My instinct was to get in the car and drive in, but that made no sense. Getting over the bridge would have been impossible. I resolved to head in at first light in the morning. Around 10 pm I got a phone call. Ricky Kay, one of the junior medical residents, was in the hospital. He was not on call, but was in the Harkness Pavilion because his wife had given birth the day before to their first daughter. So at least phone lines were still working. He and his wife were on the ninth floor and all was dark. His wife was fine but his daughter had mild infantile jaundice that required phototherapy. That was not happening without power. The Presbyterian Hospital neonatal ICU on the 18th floor did have power. He had followed as an obstetrical resident carried his daughter up the 9 flights of stairs to the unit where she could get the treatment. Since he was comfortable that his family was okay he asked if I wanted him to do anything. I told him to check the medical wards and the ICUs. Neither of us knew at that time if they had power to run the mechanical ventilators and monitors in the ICUs. I also asked him to check the ER to see what was happening down there. An hour later he called back with a mixed report. The ICUs had limited power, the ventilators were working. He detected no major crises on the wards. The staff seemed to be coping. His description of the ER was less encouraging. He said it looked like a scene out of Dante's Inferno.

I was able to make my way into the hospital by 7 am the next morning. Traffic was actually lighter than normal. Many commuters sensibly stayed home. We were able to get the house staff organized and continue the work of caring for the patients as best we could. Everything actually seemed calmer and slower. It was still hot. Nightfall came, still without power. But at 10:30 pm the night of July 14 the power came back and things slowly got back to the hospital norm. As the Lovin' Spoonful had sung; many a neck got dirty and gritty during those 27 "Hot town, Summer in the City" hours. It was just one more test and experience in the life of residents at the Columbia Presbyterian Hospital.

His Amygdala Just Fires

One upside to the Chief Residency role was the time spent with the Chief of the Department of Medicine. A few weeks before I began my Chief Residency year, Dr. Daniel V. Kimberg was hired to be Chairman of the Department of Medicine and Chief of the Medical Service. He was young (mid 40s) vibrant and forceful. He was a Columbia medical school graduate, had trained at the medical center, and been a junior faculty member for a few years before heading to the University of Rochester where he made his name and career as an investigative gastroenterologist specializing in the secretory function of the GI tract. So he knew the history and tradition of the medical school, the training program and the hospital. In addition, he brought the perspective of an outsider. The house staff was excited. The word was that he liked and valued residents. We felt he might inject a new energy and enthusiasm into the program. The past 4 years had seen

an aging and ill Chief (Dr Reagan), followed by the
interim Dr. Aranow. Both had been fine but new energy
and ideas seemed needed and were welcomed.
Throughout my year as Chief Resident I interacted
almost daily with Dr. Kimberg. It was a very interesting
and rewarding viewpoint on administrative and
academic medicine.

He had several fresh ideas for the department
and also for the house staff program. He wanted to
institute a daily morning report with the residents. This
had not been done in prior years. It seemed a good idea,
giving both he and the residents exposure to each other.
The problem from my perspective as advocate for the
house staff was the timing. There was a lot of work that
got done in those morning minutes and hours before
attending rounds at 10 am. The ward residents had to
check on the 3 interns and 30 or so patients under their
charge. All new admissions and any changes or crises
during the night had to interpreted, understood and
treated. The residents had to make sure the interns were
prepared for attending rounds. A meeting or conference
of any kind would cut into that valuable work time. Dr.
Kimberg and I agreed on a 15-30 minute meeting at 8:15
each morning. Most residents were in the house around
7 to 7:30 am if they were not on call, so this gave them
time to quickly catch up on events under their
responsibility. I think this brief meeting was all Dr.
Kimberg really wanted. He could have dictated the
parameters, but he had made me feel part of the decision
to set it up. Only the residents on the ward services and
ICUs were required to attend this morning report

exercise. Those on outpatient or the Harkness rotations were welcome, but not expected. The interns were not part of this exercise. They had too much to do. The meeting worked well for my purposes and responsibilities. It allowed me to see how well the residents were handling their responsibilities, how they responded to the pressure of the Chief's presence, and to hear their updates on the patients under their charge. Each morning I knew if there were problems I had to address with residents or interns later in the day.

The real benefit was that all the residents heard about all the patients admitted to the service. They could ask questions and offer suggestions. I basically ran the meeting, asking for a brief succinct recap of the new admissions and any significant problem or development with patients already in the hospital. Dr. Tom Morris, the associate Chief and an associate Dean in the Medical school also attended. He, Dr. Kimberg, and I offered brief suggestions and critiques. Once in awhile we got into a more detailed discussion of more interesting patients.

So this daily exercise was both functional and educational for the residents and for me. A lot of information was exchanged in succinct bites. Performance was assessed. And most importantly we all received invaluable perspective on specific cases and medicine in general from the Chief of the Department and his associate. Tough to beat that.

Dr. Kimberg had a lot going on his first year as Chief. In addition to getting to know and oversee the house staff, he was involved with a general reorganization of the faculty and the department. Many mornings he seemed preoccupied, perusing various papers on his desk while the residents, Dr. Morris and I discussed the new patients. He would look up briefly, add a pertinent comment, and resume his reading. Sometimes he would shoot me a look that said "I want you to check on that". This usually referred to a questionable statement or action by a resident. Sometimes he would look up with a long, baleful, wide eyed stare at a resident and then me. That stare was powerful. We all got the message. He did not like how that was handled and we had better fix it.

We all were more comfortable on the days he was more engaged and talkative. He had tons of clinical knowledge. When he was discussing medicine we were less likely to get the Stare. Not often, but every once in awhile, we got more than the Stare when he was really displeased. Repeat errors, inattention to detail, or a silly oversight could trigger a sudden volcanic outburst. His eyes grew wide and his strong, deep voice erupted with passion, a few choice expletives, and righteous indignation. The room would be brought to a standstill, the air sucked out. He would quickly calm, make a pertinent comment on the case and then get back to his reading. But, Oh Baby; none of us wanted to be on the receiving end of those explosions. One morning after one of those eruptions I walked out with Ricky Kay.

"You know, Jeff, it's weird how those outbursts come out of nowhere. I think he hears something that bothers him and boom, his amygdala just fires."

I chuckled a bit. That was a typical Ricky Kay observation; insightful, anatomical, and accurate. He was one smart resident and often had a humorous take on events. The amygdala is a small almond shaped area of the brain, deep in the medial temporal lobe. It plays a key role in the processing of emotions. We do not have conscious control over it. It just fires. Well, it could be, I thought. Anyway those outbursts sent a message and added to the growing aura of the Chief. Amygdala outburst or not, each and every one of us learned a whole lot from those resident morning report sessions with Dr. Kimberg.

Team Rounds

Medical grand rounds at CPMC had a tradition and style of their own. They were called Team Rounds. I guess because the whole medical "team" from the Chief of the Department down through professors, young attendings, residents, and lowly interns all participated. They were held each Thursday at 4:30 pm. Part of the tradition was that shortly after 4 pm urns of tea and plates of cookies would appear on a table in the hallway outside of the Chief of Medicine's office. This was in a wide corridor that connected the 8th floor of Presbyterian Hospital to a floor in the main tower of the medical school. The corridor was called 8 Stem.

Those urns of tea and plates of cookies were prepared each week by the department receptionist, Mrs. H. She was a sweet and kindly lady who always seemed to have a warm smile for weary interns and residents. Chit chat and gossip would be exchanged as the house staff and faculty sipped tea and rubbed shoulders for a few minutes before heading over to the formal conference. Most of us enjoyed this part of the Team Rounds tradition. I sure hope it still continues to this day.

Academic medical departments at teaching hospitals have a number of conferences per week. I have noted our Tuesday noon house staff conference. The subspecialty departments like Cardiology, GI, Renal, etc. also have conferences. But the conference with the broadest academic importance is usually called Medical Grand Rounds. A faculty attending, or at times a noted professor from another medical school, will speak for 45 – 50 minutes on a current medical topic or disease. The medical school professors, the hospital teaching staff, and the residents and interns were expected to attend. After the presentation questions and discussion usually ensued. Theoretically, all, particularly the interns and residents, are enlightened.

Another part of the CPMC Team Rounds tradition was that each conference was begun with a presentation of a patient case by a resident. The case and resident were selected by the Chief Resident. The case had to be of interest and pertinent to the announced topic of the speaker. A good speaker would refer to and make use of the case as he discussed the topic. There were expectations to the presentation of the case. It had to be brief, 3-4 minutes at most, but inclusive of all the pertinent clinical data and the working diagnosis. The presenting resident was in the spotlight. The eyes, ears, and brains of the Chief of Medicine, the whole teaching faculty, and many of his/her peers were focused on that presentation. One could make a good impression, or one could not. The setting added to the pressure. It could be intimidating. Team Rounds were held in a large medical school amphitheater. The speaker's podium was in the pit of the amphitheater with the chairs and audience extending upward for two floors. The Chief of the Department had a designated seat in the first row just in front and to the right of the presenter. I had presented a few cases as a resident in the two prior years and had fared reasonably well. Now as the Chief Medical Resident I had to select the presenters; and some of them needed a little coaching.

The Chief Medical Resident was also responsible for selecting the topics and speakers for Team Rounds. The idea was to schedule a broad range of topics and mix in junior faculty, older professors and guest speakers. Once the topic and speaker were selected and scheduled you had to find an appropriate case from the ward services to kick off the discussion. Fortunately I got some help from the Chief of the Department, Dr. Kimberg, in selecting topics and speakers. This was another area where exposure to him added to the experience of the Chief Residency year.

Throughout the ensuing 35 years since my days at CPMC I have attended many medical grand rounds at several other hospitals. It is probably nostalgia but none seemed as "grand" as those old CPMC Team Rounds. The tea, cookies, and Mrs.H probably had something to do with that.

The Red Death is no More

The Chief Medical Residency role did come with a few perks. I had an office on 8 Stem not far from the Chief of Medicine's office. I also had a secretary. She typed and distributed the house staff schedules, the conference schedules, and other assignments I prepared. As the year went on my confidence grew and I began to try to affect policies that pertained to the house staff. The Chief Resident was advocate as well as organizer for the house staff. Just as I had done two years before, the residents on the private patient Harkness Pavilion rotation continued to complain bitterly about having to prepare and administer chemotherapy treatments on private cancer patients during nights on call. This was tedious, time consuming work but also unsafe on many levels. The hospital administration had been promising to create proper 24 hour pharmacologic staffing for this role for over a year. It had not happened.

I had discussed the issue with Dr. Kimberg. He agreed it needed to be changed. That was enough for me. I wrote an impassioned (a little naïve in retrospect) letter to the hospital administrator responsible for these issues. I had sent a copy to Dr. Kimberg. A day later he appeared at my office door with the letter in hand. I was not sure what his reaction was going to be.

"Did you check with me on this?"

"Well, not exactly. But we did discuss it and you agreed it should be changed."

He gave me a wink and a little smile; then he walked out. Whew! He had my back. The next day the administrator called me to discuss the issue. Within a week the hospital had hired the proper personnel to handle chemotherapy treatments in the nighttime hours in the Harkness Pavilion and the rest of the hospital. The residents were relieved of drawing up and administering dosages of the Red Death. And more importantly, the patients were better served.

The Sheriff Redux

In late June of 1978 another transition loomed. This would be a big one for me. I was leaving the CPMC and on July 1, 1978 would start a cardiology fellowship program at Mt. Sinai Hospital in New York City. Starting with medical school I had spent 8 years in the buildings and towers of the CPMC complex. Each July 1st of those 8 years my status and responsibilities changed, but I still remained within an institution and culture I knew. It had been a long and intense road, particularly the last 4 years. I knew I would spend 2 years in the Mt. Sinai cardiology program but did not know what I would do, or where I would go after that. Coming back to CPMC as a junior cardiology attending was an option. I had been very impressed by Dr. Kimberg and the direction he was taking the medical department. I certainly would not mind being part of that, and I knew he liked my work as the Chief Medical Resident. But there were no guarantees and I had to be open to all possibilities.

On the afternoon of June 30 I had handed over the Chief Residency reins (and pager) to my successor. That morning Dr. Kimberg and I had our last meeting and said our goodbyes. He gave me two books on photography, one the western landscape photos of Ansell Adams. He knew I was interested in photography. I still have those books.

Somehow one last social gathering got organized. My good friend and the house staff social catalyst, Peggy Fracaro would not pass up the opportunity to put together a party. Adele Tedeschi joined Peggy in the set up. After the 3 years as fellow intern and resident Adele had stayed at the medical center in a junior teaching position the year I was Chief. So at 5 pm on June 30, my last official day at CPMC, I made one last trip up the elevator to Club 22. Where else could such an event occur? The view of the Big City and the Hudson River were as always, spectacular. Wine and various nibbles were consumed. I believe Mrs. H sent a plate of her famous Team Rounds cookies. Many of the residents and a few of the medical service nurses were there. A few professors and attendings dropped by. Glenda Garvey shared a glass of wine and her good wishes. Harold Neu joined the festivities, his impish grin a little the wider from the wine.

I was enjoying the moment, basking a little, proud of the last 4 years, but also having bittersweet thoughts that change and a journey's end can engender. The festivities had been going on for about 30 minutes when I noticed Dr. Whitlock, the Sheriff, leaning against the wall off by himself. He had a glass of wine in hand. He gave me a nod and a wry little grin. I was very pleased he had showed up. I walked over and we shook hands. I had spent time with him on the GI service as a senior resident. He and I had met a few times about some interns having problems this past year when I was Chief. He was still prowling the 8 West service in the early am, checking on things. I now knew he had the best interest of the interns and residents in mind, despite his gruffness and imposing demeanor. I no longer feared him as I did those mornings on 8 West 4 years ago. In fact I liked him a great deal.

"Well, Doctor Blake. You have done a pretty fair job here. Don't let up now. And think about maybe coming back here in a few years. We can use your type."

Then, not unlike those mornings on 8 West 4 years ago, he was gone. The diction was still clipped, the manner still a little brusque, and he had mixed compliment with admonition. One thing was different. In that internship year I was addressed as Blake. The Sheriff just now had put the title Doctor in front of my name. And I could tell he meant to convey respect and a bit of congratulation. Coming from him that was high praise.

Reflecting on the last 4 years I was amazed how far I had come. I too, felt comfortable calling myself a Doctor. It had been hard earned.

Notes and Acknowledgements:

Dr. Harold Neu died prematurely in 1998 at age 63. He was a renowned international figure in infectious diseases and microbiology. His name is used with the permission of his surviving (and still actively working spouse) Dr. Carmen Neu. As noted in the text, Dr. Glenda Garvey died in 2004 at age 61. Her achievements and renown are noted there as well. Her name is used with the permission of her brother, Dr. Thomas Garvey III. I thank both Carmen Neu and Thomas Garvey for their generous support of this effort.

Dr. Daniel Kimberg died suddenly and prematurely the year after my chief residency year. He was in his 40s. I suspect his tenure as Chief of Medicine would have been long and successful had he lived. Attempts were made to contact a surviving relative but were unsuccessful. I certainly hope that his relatives have no objection to the references to him in this book. I had the utmost respect for him.

Dr. Robert Whitlock continued many more years at CPMC, both as teaching faculty and a gastroenterologist. For several years he was the Chief of Medicine at the Allen Pavilion. He retired a number of years ago to the eastern shore of Maryland. His name is used with his permission. His health is not the best, but he reports that he still has that six pointed star and can still cuss pretty well. The Sheriff still looms large.

The following peers and colleagues added insight and information to the stories in the text. They

also gave permission for the use of their names. I thank
them all for their support and enthusiasm. Ricky Kay is
a cardiologist in Westchester County. He and I were
colleagues for many years and share many memories
post our Columbia years. He has come full circle,
recently re affiliating with Columbia Presbyterian.
Adele Tedeschi is retired and living in New Jersey and
Florida. She had a very successful career in internal
medicine. She tells me she still feels guilty about
saddling me with her 9 West rotation when she was out
with hepatitis. Her husband Rich Mattern continues to
work as a radiologist and remains active in the alumni
affairs of our medical school class. He misses Club 22.
The amazing Peggy Fracaro continues to work at CPMC
where she is the Nursing Director of Infection
Prevention and Control. I suspect she continues to
monitor the social network of the hospital. She refuses
to acknowledge whether she still visits Club 22. Joe
Tenenbaum is a cardiologist and continues to work at
CPMC as the Chief of Medicine at the Allen Pavilion
Campus. He is also the head of the house staff training
program.

All patient names used are fictional except that
of Charles Lindbergh. He was a public figure and use of
his real name was felt to be instrumental to the particular
story where it is used. Since he was a public figure and
died many years ago in 1974, I truly think the use of his
name is not a violation of his privacy. All other
descriptions of patients are based on real patients and
real incidents, but their real names are not used. Some

patient descriptions represent composites of a few different patients.

This is a memoir based on my recollections. These recollections are supported in many cases by the colleagues noted above. All statements in quotations attributed to real people are actually my recollections of what might have been said. They are not meant or guaranteed to be actual statements by the speakers. The quotations are used to make the relating of the stories better.

My career path did not take me back to CPMC after I left in 1978. My practice in cardiology was in Westchester County, New York and my academic affiliation with New York Medical College. I am now retired but do part time work as a volunteer physician in clinics for uninsured and underinsured patients. I continue to learn how to be a better doctor.

In 1974 I was one of 16 freshly minted MDs, aspiring to be doctors. As best I know all of us had successful careers in medicine. My hat is off to all of us. I know that I truly value the training I received at CPMC. I would wager that all those former colleagues feel the same way.